Sour Water: Should We Be Worried?

Harper Nomad

1st edition 2024

Published by Manley Publications

Understanding PFAS – The Basics

What They Are and Their Chemical Properties

Per- and Polyfluoroalkyl Substances, commonly known as PFAS, represent a vast family of man-made chemicals known for their unique properties. These compounds have become an increasingly prominent topic of concern due to their persistence and potential hazards in the environment. At their core, PFAS are characterised by the presence of carbon-fluorine bonds, which are among the strongest bonds in organic chemistry. This strength imparts remarkable stability to PFAS molecules, making them resistant to degradation.

PFAS include a range of chemicals, such as Perfluorooctanoic Acid (PFOA), Perfluorooctane Sulfonate (PFOS), and thousands of other related substances. Each variant exhibits its own set of chemical properties, but they all share the common feature of being highly resistant to both water and oil. This resistance stems from the fluorine atoms bonded to the carbon chain, creating a barrier that repels liquids and prevents the breakdown of the compound.

The chemical structure of PFAS typically features a carbon chain fully or partially fluorinated, which endows the substance with its oil- and water-repellent properties. This unique structure makes PFAS incredibly useful in a wide range of applications. However, it also contributes to their persistence in the environment and in human bodies. Unlike most organic compounds that eventually break down through natural processes, PFAS resist degradation due to their strong carbon-fluorine bonds.

The stability of PFAS molecules is what earns them the nickname "forever chemicals". Once released into the environment, these substances do not break down easily and can remain in soil, water, and living organisms for decades or even centuries. Their durability means that they can accumulate over time, leading to widespread contamination. This persistence poses significant challenges for environmental management and public health.

Understanding the fundamental properties of PFAS is crucial to comprehending the broader implications of their use and contamination. The same characteristics that make PFAS valuable for various industrial applications—such as their resistance to heat, stains, and water—are the very traits that also make them problematic from an environmental and health perspective. Their persistence and accumulation lead to long-term contamination issues that are difficult to remediate.

In summary, PFAS are a diverse group of synthetic chemicals with unique chemical properties due to their strong carbon-fluorine bonds. Their resistance to environmental degradation and their ability to repel water and oil have led to widespread use in various industries. However, these same properties contribute to their persistence in the environment, making them a significant concern for both environmental and public health.

Historical Use of PFAS in Various Industries

The utilisation of Per- and Polyfluoroalkyl Substances (PFAS) in various industries reflects their exceptional chemical properties, which have rendered them indispensable in a range of applications. From firefighting foams to non-stick cookware, PFAS have played a significant role in modern industrial practices, but their historical use also sheds light on how pervasive these chemicals have become in our environment.

Firefighting Foams:

One of the earliest and most prominent uses of PFAS was in firefighting foams, specifically Aqueous Film-Forming Foam (AFFF). Developed in the 1960s, AFFF contains PFAS compounds due to their ability to quickly suppress flammable liquid fires, such as those involving aviation fuel or petroleum products. PFAS-based foams form a thin, protective film that spreads across the surface of burning liquids, smothering the flames and preventing the release of flammable vapours. This effectiveness in emergency situations made AFFF a standard tool in both military and civilian firefighting practices.

The widespread use of AFFF, particularly at airports, military bases, and industrial sites, has led to substantial environmental contamination. When used, these foams often leach into soil and groundwater, contributing to the persistence of PFAS in these environments. Despite their efficacy in fire suppression, the environmental impact of PFAS in firefighting foams has become a significant concern as contamination has spread to nearby water sources and communities.

Non-Stick Cookware:

Another major application of PFAS has been in the production of non-stick coatings for cookware. Polytetrafluoroethylene (PTFE), commonly known by the brand name Teflon, is a PFAS compound known for its remarkable non-stick properties. Introduced in the 1940s, PTFE became a popular choice for coating frying pans, baking trays, and other kitchen utensils due to its ability to repel grease and resist high temperatures.

The popularity of non-stick cookware expanded rapidly, leading to widespread use in households and commercial kitchens. While the convenience of easy-to-clean cookware has been appreciated by many, the environmental and health implications of PFAS in these products have become a growing concern. During the manufacturing process, as well as through the use and disposal of non-stick cookware, PFAS can be released into the environment, contributing to contamination.

Stain-Resistant Fabrics:

PFAS compounds have also been employed in the textile industry to produce stain-resistant and water-repellent fabrics. Fabrics treated with PFAS-based coatings resist staining from liquids such as coffee or wine, and they remain clean and dry for extended periods. This application proved particularly advantageous for carpets, upholstery, and outdoor gear, where durability and resistance to environmental factors are crucial.

The use of PFAS in textiles dates back several decades, driven by consumer demand for low-maintenance products. However, as with other PFAS applications, the environmental impact of these chemicals has come under scrutiny. The treatment of textiles with PFAS leads to the release of these substances into the environment through washing and disposal, further contributing to the accumulation of PFAS in soil and water sources.

Industrial Applications:

In addition to consumer products, PFAS have found extensive use in various industrial applications due to their chemical stability and resistance to heat and corrosion. They have been used in the manufacturing of semiconductors, electroplating, and in certain chemical processes where high temperatures and corrosive conditions are present. Their ability to withstand harsh conditions has made PFAS a valuable asset in industries requiring reliable performance under challenging environments.

The industrial use of PFAS often involves the release of these chemicals into the environment, either through emissions, waste disposal, or leaks. Over time, the accumulation of PFAS from industrial activities has led to widespread contamination, affecting both the local environment and broader water systems.

Regulatory Oversight and Historical Context:

Historically, the potential environmental and health risks associated with PFAS were not fully understood or regulated. The primary focus was on their functional benefits rather than their long-term impacts. As scientific research advanced and concerns about PFAS contamination grew, regulatory bodies began to address the issue more actively. However, the legacy of PFAS use continues to affect communities and ecosystems around the world.

In summary, PFAS have been widely used across various industries due to their unique chemical properties. Their applications in firefighting foams, non-stick cookware, stain-resistant fabrics, and industrial processes illustrate their versatility and effectiveness. However, the historical use of PFAS has also led to significant environmental and health concerns, highlighting the need for ongoing efforts to manage and mitigate their impact. Understanding the historical context of PFAS use is crucial for addressing the challenges posed by these persistent chemicals.

The Term "Forever Chemicals" – Why PFAS Are Persistent

The term "forever chemicals" is often used to describe Per- and Polyfluoroalkyl Substances (PFAS), and it aptly reflects the persistent nature of these substances in the environment. To understand why PFAS have earned this moniker, it is essential to delve into the chemical properties that contribute to their longevity, the processes by which they accumulate, and the broader implications of their persistence.

Chemical Stability and Resistance:

At the heart of PFAS's persistence lies their unique chemical structure. PFAS molecules feature carbon-fluorine bonds, which are exceptionally strong compared to other chemical bonds. The carbon-fluorine bond is one of the strongest bonds in organic chemistry, providing PFAS with remarkable stability and resistance to chemical breakdown. This stability is due to the high electronegativity of fluorine atoms, which create a tight bond with carbon, effectively shielding the molecule from degradation.

Unlike many organic compounds that are susceptible to natural processes such as oxidation or hydrolysis, PFAS do not easily break down when exposed to environmental factors. This resistance to chemical degradation means that once PFAS enter the environment, they tend to remain there for extended periods, accumulating over time and posing long-term challenges for contamination control.

Environmental Accumulation:

PFAS enter the environment through various pathways, including industrial discharges, leachate from landfills, and the use of products containing these chemicals. Once released, PFAS can migrate through soil and water, spreading contamination over large areas. Their persistence in the environment means that even after the source of contamination is removed or mitigated, PFAS can continue to affect the ecosystem for decades.

Water systems are particularly susceptible to PFAS contamination due to the chemicals' high solubility and resistance to degradation. PFAS can accumulate in rivers, lakes, and groundwater, leading to widespread contamination of water supplies. The movement of PFAS through water systems can also lead to their entry into the food chain, as aquatic organisms absorb these chemicals, which can then be consumed by humans and animals.

Bioaccumulation and Biomagnification:

One of the most concerning aspects of PFAS persistence is their ability to bioaccumulate and biomagnify within living organisms. Bioaccumulation refers to the process by which organisms accumulate PFAS in their tissues over time, often at levels higher than those found in the surrounding environment. This occurs because PFAS are not easily metabolised or excreted by organisms, leading to their build-up in fat tissues.

Biomagnification is a related phenomenon where PFAS concentrations increase as they move up the food chain. For example, small aquatic organisms that have absorbed PFAS from their environment are consumed by larger fish, which in turn are eaten by predators, including humans. This process can lead to significantly higher levels of PFAS in top predators compared to those in the surrounding environment.

Challenges in Remediation:

The persistence of PFAS presents significant challenges for remediation efforts. Traditional methods of cleaning up environmental contamination, such as chemical treatment, physical removal, or biological degradation, are often ineffective against PFAS due to their stability and resistance to breakdown. As a result, specialised techniques are required to address PFAS contamination, including advanced oxidation processes, activated carbon adsorption, and ion exchange.

However, these methods are often costly and may not completely eliminate PFAS from the environment. Additionally, the complex nature of PFAS contamination requires ongoing monitoring and management to ensure that remediation efforts are effective and that new sources of contamination are addressed.

Long-Term Implications:

The persistence of PFAS has far-reaching implications for environmental health and human safety. Contaminated water supplies can lead to exposure risks for communities, while the accumulation of PFAS in food sources can impact public health. The long-term environmental effects of PFAS, including their potential impact on biodiversity and ecosystem function, also pose significant concerns.

Addressing PFAS contamination requires a multifaceted approach that includes regulation, monitoring, and innovative remediation technologies. It also necessitates a commitment to reducing the use of PFAS in products and processes to prevent further environmental contamination.

Conclusion:

In conclusion, PFAS are termed "forever chemicals" due to their exceptional chemical stability and resistance to environmental degradation. Their ability to persist in the environment, accumulate in living organisms, and biomagnify through the food chain underscores the challenges associated with managing PFAS contamination. Understanding the reasons behind PFAS persistence is crucial for developing effective strategies to address their impact on the environment and public health. As the global community continues to grapple with the consequences of PFAS pollution, recognising the enduring nature of these chemicals will be essential in crafting solutions and mitigating their effects.

Basic Science Behind PFAS Accumulation in Water Sources

The accumulation of Per- and Polyfluoroalkyl Substances (PFAS) in water sources is a complex issue driven by the unique chemical properties of these substances, their environmental behaviour, and the interactions with water systems. To grasp the full extent of PFAS contamination, it is essential to understand the fundamental science behind how these chemicals accumulate in water, the processes involved, and the challenges posed by their persistence.

Chemical Properties and Solubility:

PFAS are characterised by their carbon-fluorine bonds, which are among the strongest in organic chemistry. This strong bond imparts a high degree of chemical stability to PFAS, making them resistant to decomposition by environmental processes. In aqueous environments, PFAS exhibit significant solubility due to their polar nature. The fluorinated carbon chain creates a strong repulsion to water molecules, yet the carboxyl or sulfonic acid groups in many PFAS compounds can form weak hydrogen bonds with water. This dual nature allows PFAS to dissolve readily in water, facilitating their widespread distribution.

The solubility of PFAS in water means that once they are released into aquatic environments, they can disperse over large areas, making them difficult to contain or remediate. This high solubility also means that PFAS can remain in water systems for extended periods, contributing to long-term contamination issues.

Transport and Migration in Aquatic Systems:

Once PFAS enter water sources, their movement through the environment is influenced by several factors, including water flow, adsorption to sediments, and interactions with other chemicals. In rivers, lakes, and groundwater, PFAS are transported by the flow of water, which can carry them over long distances from their original source. This transport mechanism can lead to widespread contamination, affecting areas far from the initial point of release.

PFAS can also interact with particulate matter in water, such as sediments and organic matter. While some PFAS may adhere to sediments, the strong carbon-fluorine bonds make it challenging for them to bind firmly. As a result, PFAS can remain in the water column and continue to migrate through aquatic systems. Additionally, PFAS can accumulate in the sediment, where they may persist for extended periods, potentially resuspending into the water column under certain conditions.

Biological Uptake and Bioaccumulation:

One of the critical factors in PFAS accumulation is their uptake by aquatic organisms. PFAS can be absorbed by fish and other aquatic life through their gills, skin, or ingestion of contaminated water and food. Once inside the organism, PFAS tend to accumulate in fat tissues due to their lipophilic nature. Unlike many other contaminants, PFAS are not easily metabolised or excreted, leading to a gradual build-up within the organism's tissues.

Bioaccumulation occurs when an organism absorbs a higher concentration of PFAS from its environment than it can eliminate. This accumulation is particularly concerning in the case of species that are high on the food chain. As these organisms are consumed by predators, including humans, PFAS concentrations can increase up the food chain—a phenomenon known as biomagnification. This process can result in significantly higher PFAS levels in top predators compared to those in the surrounding environment.

Remediation and Treatment Challenges:

The persistence of PFAS in water sources presents significant challenges for remediation and treatment. Traditional water treatment methods, such as coagulation, sedimentation, and conventional filtration, are often ineffective at removing PFAS due to their chemical stability and solubility. As a result, specialised treatment technologies are required to address PFAS contamination.

Activated carbon adsorption is one of the most commonly used methods for PFAS removal. In this process, water is passed through activated carbon filters, which adsorb PFAS from the water. While effective, activated carbon needs frequent replacement or regeneration, and it may not remove all PFAS compounds, particularly those with shorter carbon chains.

Ion exchange resins are another treatment option, which involves exchanging ions in the water with ions on the resin. This method can be effective for removing certain PFAS, but like activated carbon, it requires regular maintenance and may not be universally applicable.

Advanced oxidation processes (AOPs) utilise powerful oxidants, such as ozone or hydrogen peroxide, to break down PFAS molecules. AOPs can degrade PFAS into less harmful substances, but they can be costly and complex to implement on a large scale.

Regulatory and Monitoring Challenges:

The detection and monitoring of PFAS in water sources also present significant challenges. PFAS compounds are often present at very low concentrations, requiring highly sensitive analytical methods for accurate measurement. Techniques such as liquid chromatography coupled with mass spectrometry (LC-MS) are commonly used to detect and quantify PFAS, but these methods are expensive and require specialised equipment and expertise.

Regulatory frameworks for PFAS contamination vary widely, with some regions implementing stringent limits and others lagging in enforcement. In the UK, the regulatory focus has been increasing, with agencies setting limits for PFAS in drinking water and requiring monitoring and reporting. However, the wide range of PFAS compounds and their diverse sources make comprehensive regulation and enforcement challenging.

Future Directions and Research Needs:

Addressing PFAS contamination in water sources requires ongoing research and innovation. Future research should focus on developing more effective and cost-efficient treatment technologies, improving monitoring and detection methods, and understanding the long-term impacts of PFAS on ecosystems and human health. Advancements in science and technology will be crucial in managing and mitigating PFAS contamination and reducing their environmental and health impacts.

Conclusion:

The accumulation of PFAS in water sources is a complex issue driven by their chemical properties, environmental behaviour, and interactions with aquatic systems. Their solubility, persistence, and ability to bioaccumulate and biomagnify make PFAS challenging to manage and remediate. Understanding the basic science behind PFAS accumulation is essential for developing effective strategies to address contamination and mitigate its impacts. As research and technology advance, it is crucial to continue addressing the challenges posed by PFAS to protect water resources and public health.

The Journey of PFAS – From Production to Pollution

The Lifecycle of PFAS – From Industrial Production to Environmental Release

Understanding the lifecycle of Per- and Polyfluoroalkyl Substances (PFAS) provides crucial insight into how these persistent chemicals transition from industrial production to widespread environmental contamination. This journey involves several stages, each contributing to the eventual presence of PFAS in water systems, soils, and living organisms.

1. Production and Manufacturing:

The lifecycle of PFAS begins with their production and manufacturing. The synthesis of PFAS involves complex chemical processes that typically include the creation of fluorinated carbon chains. The production of PFAS starts with the fluorination of hydrocarbons or fluorinated precursors. This process is carried out in specialised chemical facilities where fluorine gas reacts with hydrocarbons to produce various PFAS compounds.

In the mid-20th century, PFAS were initially manufactured to serve a range of industrial applications due to their unique chemical properties. The first commercially available PFAS were used in products like Teflon, which is a type of polytetrafluoroethylene (PTFE). Over time, the use of PFAS expanded to include applications in firefighting foams, stain-resistant coatings, and industrial processes.

The production of PFAS generates large quantities of by-products and waste materials. These by-products can include various PFAS-related chemicals, which are sometimes discarded or inadequately treated. The handling and disposal of these by-products pose significant environmental risks, as they can contribute to PFAS contamination.

2. Use in Consumer and Industrial Products:

Once manufactured, PFAS are incorporated into a wide array of consumer and industrial products. Their application in products like non-stick cookware, stain-resistant fabrics, and water-repellent treatments is widespread. In these products, PFAS are used to impart desirable properties such as resistance to heat, stains, and water.

In consumer products, PFAS compounds are often present in relatively high concentrations, and their release into the environment can occur through several mechanisms. For example, when non-stick cookware is used, PFAS can be released into the air or food through degradation of the coating. Similarly, stain-resistant treatments on carpets and upholstery can lead to PFAS shedding into the environment during normal use and cleaning.

In industrial settings, PFAS are employed in processes requiring heat resistance or chemical stability. For instance, they are used in the production of semiconductors, metal plating, and as additives in various chemical processes. During these industrial activities, PFAS can be released into the environment through emissions, spills, and improper disposal of waste products.

3. Disposal and Waste Management:

The disposal and waste management of PFAS-containing products and by-products pose significant environmental challenges. PFAS are not easily broken down by conventional waste treatment methods, leading to the accumulation of these substances in landfills and other disposal sites.

In landfills, PFAS can leach into the surrounding soil and groundwater, particularly if the landfill liner or containment system is compromised. Leachate from landfills containing PFAS can eventually enter local water systems, leading to widespread contamination. The persistence of PFAS means that they can remain in landfill leachate for decades, continually contributing to environmental pollution.

Similarly, waste incineration, which is often used to dispose of PFAS-containing products, can lead to the release of PFAS into the atmosphere. Although high-temperature incineration can break down some chemical compounds, PFAS are resistant to thermal degradation, resulting in their release into the air or formation of hazardous residues.

4. Environmental Release and Contamination:

The transition from industrial production and use to environmental contamination involves several pathways. The primary routes through which PFAS enter water systems include industrial discharges, runoff from contaminated sites, and accidental releases.

Industrial Discharges:

Manufacturing facilities that produce or use PFAS often discharge wastewater containing these chemicals into local water bodies. Despite regulations aimed at controlling industrial discharge, PFAS can still be released in significant quantities, particularly if proper treatment and containment measures are not in place. The treated wastewater may still contain trace amounts of PFAS, which can contribute to contamination in receiving water bodies.

Runoff from Contaminated Sites:

Areas where PFAS-containing products were used or disposed of can become sources of contamination through runoff. For example, airports and military bases that used PFAS-based firefighting foams may have significant soil and water contamination due to the leaching of PFAS into groundwater and surface water. Rainfall and surface runoff from these contaminated sites can carry PFAS into nearby water bodies, leading to broader contamination.

Accidental Releases:

Accidental releases of PFAS can occur during transportation, handling, or storage of PFAS-containing materials. Spills or leaks can introduce PFAS into soil and water, where they can spread and contribute to contamination. The long-lasting nature of PFAS means that such accidental releases can have persistent effects on the environment.

5. Long-Term Persistence and Accumulation:

The final stage in the lifecycle of PFAS is their long-term persistence and accumulation in the environment. Due to their chemical stability, PFAS do not readily degrade, leading to their accumulation in soil, water, and sediments. Over time, this accumulation can result in widespread contamination that affects ecosystems and human health.

PFAS can persist in water sources, where they can be taken up by aquatic organisms and enter the food chain. This bioaccumulation and biomagnification lead to higher concentrations of PFAS in top predators, including fish and birds, and potentially in humans who consume contaminated food and water.

The widespread environmental presence of PFAS necessitates ongoing monitoring, regulation, and remediation efforts to address contamination and mitigate its impacts. Understanding the lifecycle of PFAS is crucial for developing effective strategies to manage their release, minimise environmental contamination, and protect public health.

Conclusion:

The journey of PFAS from production to pollution involves several critical stages, including industrial production, use in consumer and industrial products, disposal and waste management, and eventual environmental release. Each stage contributes to the potential for PFAS to enter and contaminate water systems and other environmental media. Recognising these pathways and understanding the persistence of PFAS in the environment are essential for addressing the challenges posed by these persistent chemicals and developing effective strategies for remediation and prevention.

Common Pathways Through Which PFAS Enter Water Systems

Per- and Polyfluoroalkyl Substances (PFAS) are notorious for their ability to infiltrate and persist in water systems due to their chemical properties and widespread use. Understanding the various pathways through which PFAS enter aquatic environments is crucial for identifying sources of contamination and developing effective mitigation strategies. This section examines the common routes of PFAS entry into water systems, including industrial discharges, landfill leachate, wastewater treatment plants, and stormwater runoff.

1. Industrial Discharges:

Industrial discharges are a primary pathway through which PFAS enter water systems. Facilities involved in the production or use of PFAS-containing products often release wastewater containing these chemicals into local water bodies. These discharges can occur directly into rivers, lakes, or seas, or into municipal sewage systems, which eventually lead to surface waters.

Manufacturing Facilities:

Industries that manufacture PFAS or use them in their processes, such as those producing non-stick coatings, firefighting foams, or industrial lubricants, can be significant sources of PFAS contamination. These facilities often have wastewater containing PFAS as a by-product of their operations. Despite treatment processes designed to remove contaminants, PFAS can remain in the treated effluent due to their chemical stability and resistance to conventional treatment methods.

Processing Facilities:

Facilities that handle PFAS-containing products, such as textile or coating industries, may also contribute to contamination. These facilities can generate waste containing PFAS through various means, including spills, leaks, and process water. When this waste is not properly managed or treated, PFAS can enter nearby water systems, contributing to broader contamination.

2. Landfill Leachate:

Landfills where PFAS-containing products have been disposed of can become significant sources of PFAS contamination. The leachate, which is the liquid that percolates through the landfill and collects contaminants, can contain high concentrations of PFAS. Over time, the PFAS in the leachate can migrate into groundwater and surface water systems.

Landfill Sites:

Landfills that accept waste containing PFAS, such as old firefighting foams, non-stick cookware, or stain-resistant textiles, can be a major source of contamination. The PFAS in these materials can leach into the surrounding soil and groundwater, particularly if the landfill liner or containment system is compromised. This leachate can then flow into nearby water bodies, contributing to contamination.

Old or Unregulated Landfills:

Historically, many landfills operated without stringent environmental controls, leading to the release of PFAS into the environment. These unregulated or improperly managed landfills can still be sources of PFAS contamination, as the chemicals continue to leach into groundwater and surface water long after the landfill has ceased operation.

3. Wastewater Treatment Plants:

Wastewater treatment plants play a crucial role in managing and treating municipal and industrial waste. However, they can also be pathways for PFAS entry into water systems. PFAS present in influent wastewater can persist through the treatment process and end up in the treated effluent or the biosolids produced by the plant.

Municipal Wastewater:

Municipal wastewater, which includes sewage from households and businesses, can contain PFAS from various sources such as cleaning products, food packaging, and personal care items. Although wastewater treatment plants are designed to remove many contaminants, PFAS are resistant to conventional treatment methods, meaning that these substances can pass through the plant and be discharged into surface waters.

Industrial Wastewater:

Industrial wastewater often contains higher concentrations of PFAS compared to municipal wastewater. Industries that use PFAS in their processes or dispose of PFAS-containing materials can contribute significant amounts of these chemicals to wastewater treatment plants. The treatment processes may not be sufficient to remove PFAS, leading to their discharge into the environment.

4. Stormwater Runoff:

Stormwater runoff is another significant pathway for PFAS to enter water systems. During rainfall or snowmelt, water can pick up pollutants from various surfaces and carry them into storm drains, which eventually discharge into rivers, lakes, or seas.

Urban Areas:

In urban areas, stormwater runoff can pick up PFAS from surfaces such as roads, roofs, and pavements. For example, PFAS-containing products like stain-resistant treatments for carpets or waterproofing for textiles may contribute to runoff contamination when these materials are washed or degraded by rainfall.

Industrial Sites:

Stormwater runoff from industrial sites where PFAS are used or stored can also contribute to contamination. Areas where PFAS-containing materials are handled or disposed of may have

PFAS residues on surfaces that can be washed away by rain. This runoff can then carry PFAS into stormwater systems and eventually into natural water bodies.

5. Accidental Releases and Spills:

Accidental releases and spills of PFAS-containing substances can lead to significant contamination of water systems. Such incidents can occur during the transportation, storage, or handling of PFAS materials, and can result in the direct release of PFAS into the environment.

Transportation Accidents:

Transportation accidents involving PFAS-containing products or chemicals can lead to spills and leaks. For example, a truck carrying firefighting foam or industrial chemicals may overturn, releasing PFAS into the soil and nearby water bodies. The resulting contamination can be difficult to contain and remediate, leading to long-term environmental impact.

Storage and Handling Issues:

Improper storage or handling of PFAS-containing materials can also lead to accidental releases. For instance, leaks from storage tanks or containers can result in PFAS contamination of soil and groundwater. These incidents can be exacerbated by inadequate containment measures or maintenance practices.

6. Historical Contamination:

Past industrial practices and waste disposal methods have led to historical PFAS contamination that continues to affect water systems. Sites with a history of PFAS use or disposal may still be sources of ongoing contamination, even if current practices are more regulated.

Legacy Sites:

Legacy sites, such as old military bases, airports, and industrial facilities, often have a history of PFAS use, particularly with firefighting foams. These sites may have significant soil and groundwater contamination that continues to impact nearby water sources. Despite efforts to address contamination, the persistence of PFAS means that these legacy sites can remain sources of pollution for decades.

Conclusion:

PFAS enter water systems through multiple pathways, including industrial discharges, landfill leachate, wastewater treatment plants, stormwater runoff, and accidental releases. Each of these pathways contributes to the widespread presence of PFAS in the environment and presents challenges for managing and mitigating contamination. Understanding these pathways is essential for developing effective strategies to address PFAS pollution and protect water resources. Effective regulation, improved treatment technologies, and targeted remediation efforts are crucial in tackling the complex issue of PFAS contamination and reducing its impact on both the environment and public health.

Case Studies Highlighting Major Incidents of PFAS Contamination

To understand the full impact of Per- and Polyfluoroalkyl Substances (PFAS) on water systems, examining specific case studies of major contamination incidents provides valuable insights. These case studies illustrate the widespread nature of PFAS pollution, the challenges of addressing it, and the often severe consequences for both the environment and public health. This section explores notable PFAS contamination events, including the impacts on communities, responses to the contamination, and lessons learned.

1. The Dover Air Force Base, USA

The Dover Air Force Base in Delaware, USA, represents a significant case of PFAS contamination linked to the use of firefighting foams. These foams, which contain PFAS compounds, were used extensively at the base for fire training and emergency response exercises.

Background:

During the 1970s and 1980s, the Dover Air Force Base utilised PFAS-based firefighting foams to suppress fires and train personnel. Over time, these foams were released into the environment, either through direct application during drills or through accidental spills. The foams were discharged into storm drains, which led to the contamination of local soil and groundwater.

Contamination and Impact:

Studies revealed high concentrations of PFAS in the groundwater around the base, with levels significantly exceeding health advisory limits. The contamination extended beyond the base, affecting nearby residential areas and local water supplies. This resulted in widespread concerns among the community about potential health risks, including elevated levels of PFAS in drinking water.

Response and Remediation:

In response to the contamination, the U.S. Department of Defence initiated a comprehensive investigation and remediation programme. This included identifying and assessing the extent of PFAS contamination, providing alternative water sources to affected communities, and implementing remediation technologies such as activated carbon filtration to remove PFAS from groundwater. Despite these efforts, the long-term impacts of PFAS contamination at the site remain a concern, highlighting the challenges of effectively addressing such pervasive pollution.

2. The Tyne and Wear Fire and Rescue Service, UK

The Tyne and Wear Fire and Rescue Service (TWFRS) in the UK is another notable example of PFAS contamination stemming from firefighting practices. The use of PFAS-containing firefighting foams led to significant environmental issues in the region.

Background:

Similar to many other fire services, TWFRS used PFAS-based foams for firefighting and training exercises. Over time, these foams were discharged into training grounds and fire stations, leading to the accumulation of PFAS in soil and groundwater.

Contamination and Impact:

PFAS contamination was detected in the vicinity of several fire stations and training facilities operated by TWFRS. The contamination extended into nearby water sources, raising concerns about potential impacts on drinking water supplies and local ecosystems. Investigations revealed elevated levels of PFAS in both soil and groundwater, prompting the need for further assessment and remediation.

Response and Remediation:

TWFRS, along with environmental agencies, undertook a series of actions to address the contamination. This included detailed site investigations, monitoring of water quality, and the implementation of interim measures to mitigate risks. Remediation efforts involved treating contaminated groundwater using advanced filtration techniques and removing affected soil. The case emphasised the importance of addressing historical practices and the need for ongoing monitoring to manage and reduce PFAS contamination effectively.

3. The Shropshire Incident, UK

The Shropshire incident highlights the challenges associated with PFAS contamination from multiple sources. In this case, contamination resulted from both industrial processes and landfill sites.

Background:

In Shropshire, industrial activities involving PFAS-containing substances, combined with the use of PFAS-based products and improper disposal, led to significant contamination. Additionally, old landfills containing PFAS-containing waste contributed to the pollution.

Contamination and Impact:

PFAS contamination was detected in several locations across Shropshire, including groundwater, surface water, and soil. The contamination had potential implications for local water supplies and agricultural activities. Investigations revealed a complex network of pollution sources, including historical industrial operations and landfill leachate, contributing to the widespread contamination.

Response and Remediation:

Addressing the Shropshire contamination involved a multi-faceted approach. This included identifying and mapping contamination sources, conducting risk assessments, and implementing remediation strategies. The response required coordination between local authorities, environmental agencies, and industrial stakeholders. Remediation efforts included treating contaminated groundwater, managing landfill leachate, and conducting ongoing monitoring to track progress and ensure that contamination levels were reduced.

4. The Australian Firefighting Foam Contamination

Australia has faced significant challenges related to PFAS contamination from firefighting foams, with widespread incidents affecting various regions.

Background:

The use of PFAS-containing firefighting foams across Australia led to extensive contamination. Firefighting training exercises and emergency response activities at various locations, including airports, military bases, and fire stations, resulted in the release of PFAS into the environment.

Contamination and Impact:

PFAS contamination was detected in groundwater and surface water near several affected sites. The contamination impacted local water supplies and raised concerns about potential health risks for communities. The extensive use of PFAS-based foams across multiple sites meant that addressing contamination required a coordinated national effort.

Response and Remediation:

The Australian government and relevant authorities initiated a national response to the PFAS contamination issue. This included establishing guidelines for PFAS management, conducting comprehensive investigations, and implementing remediation measures. Efforts focused on identifying contamination sources, providing alternative water supplies, and applying treatment technologies to remove PFAS from affected water sources. The case underscores the need for robust management strategies and preventative measures to address PFAS pollution on a national scale.

5. The Hoosick Falls, New York Incident, USA

Hoosick Falls, a small village in New York, experienced severe PFAS contamination resulting from industrial activities.

Background:

In Hoosick Falls, PFAS contamination was traced back to a local manufacturing plant that used PFAS in its operations. The plant discharged PFAS-containing waste into a local water body, leading to contamination of the village's drinking water supply.

Contamination and Impact:

Residents of Hoosick Falls were exposed to elevated levels of PFAS in their drinking water, leading to widespread public health concerns. The contamination affected the local water supply, prompting investigations and public health advisories.

Response and Remediation:

In response to the contamination, local, state, and federal authorities took action to address the issue. This included providing alternative water sources to affected residents, implementing water treatment technologies, and conducting long-term health studies to assess the impact of PFAS exposure. The incident highlighted the critical need for proactive measures to prevent and manage PFAS contamination and the importance of community engagement in addressing environmental health issues.

Conclusion:

These case studies demonstrate the widespread and persistent nature of PFAS contamination, with significant impacts on water systems and public health. Each case underscores the importance of understanding the pathways of PFAS entry into the environment, the challenges of remediation, and the need for effective management strategies. Addressing PFAS contamination requires a coordinated effort involving regulatory agencies, industry stakeholders, and affected communities to mitigate risks and prevent future incidents. The lessons learned from these cases can inform future strategies for managing PFAS pollution and protecting water resources.

The Health Impacts of PFAS Contamination

Understanding the health impacts of Per- and Polyfluoroalkyl Substances (PFAS) contamination is crucial for assessing the overall risks associated with these persistent chemicals. PFAS are known for their longevity in the environment and their potential to accumulate in the human body, leading to a range of adverse health effects. This section explores the health risks associated with PFAS exposure, including their impact on various organ systems, potential links to chronic diseases, and the challenges of assessing and mitigating these health effects.

1. Bioaccumulation and Persistence:

PFAS are characterised by their resistance to degradation, which leads to their bioaccumulation in living organisms. These chemicals can accumulate in the human body, particularly in the liver and blood, due to their chemical structure, which allows them to persist over long periods.

Bioaccumulation Mechanism:

PFAS accumulate in the human body primarily through ingestion of contaminated water, food, or inhalation of airborne particles. Once absorbed, PFAS can bind to proteins in the blood and tissues, where they can remain for years. This bioaccumulation can lead to higher concentrations of PFAS in individuals who are exposed to contaminated sources over extended periods.

Long-Term Persistence:

The long half-life of PFAS compounds means that they can remain in the human body for decades. For example, some PFAS have half-lives of several years, which allows them to build up in the body over time. This long-term persistence raises concerns about potential chronic health effects, as even low-level exposure can lead to significant accumulation and health risks.

2. Health Effects on Various Organ Systems:

PFAS exposure has been linked to a range of adverse health effects, impacting various organ systems. Research has shown that PFAS can interfere with normal biological processes and contribute to the development of several health conditions.

Immune System:

PFAS exposure has been associated with immune system dysfunction. Studies have shown that PFAS can impair the immune system's ability to respond to infections and vaccines. For example, research has indicated that PFAS exposure can reduce vaccine efficacy and increase susceptibility to infectious diseases. The immune system's compromised function can lead to increased risk of illness and decreased ability to fight off infections.

Endocrine System:

PFAS have been shown to affect the endocrine system, which regulates hormone production and function. Evidence suggests that PFAS exposure can disrupt thyroid hormone levels and contribute to thyroid disease. Additionally, PFAS may interfere with the regulation of sex hormones, potentially affecting reproductive health and development.

Cardiovascular System:

There is growing evidence linking PFAS exposure to cardiovascular health issues. Studies have reported associations between PFAS and increased risk of hypertension, elevated cholesterol levels, and other cardiovascular conditions. The potential for PFAS to contribute to heart disease and stroke underscores the importance of understanding and addressing these health risks.

Cancer Risk:

Research has identified potential links between PFAS exposure and an increased risk of certain cancers. Notable among these is the association between PFAS and kidney cancer, as well as testicular cancer. Some studies have found elevated levels of PFAS in individuals with these types of cancer, suggesting a possible role of PFAS in cancer development.

3. Community Health Impacts and Public Awareness:

The health impacts of PFAS contamination extend beyond individual cases, affecting entire communities exposed to contaminated water or soil. Public awareness and community health initiatives play a crucial role in addressing the health consequences of PFAS exposure.

Community Health Studies:

In areas affected by PFAS contamination, community health studies have been conducted to assess the extent of health impacts and identify potential trends. These studies often involve monitoring levels of PFAS in the blood of residents, as well as conducting epidemiological research to evaluate the prevalence of health conditions associated with PFAS exposure.

Public Health Interventions:

Public health interventions in contaminated areas may include providing alternative sources of clean drinking water, implementing health screenings for affected populations, and promoting public education about the risks of PFAS exposure. Addressing health concerns through these measures can help mitigate the impact of PFAS contamination and support affected communities.

Regulatory Measures and Guidelines:

Regulatory agencies have established guidelines and limits for PFAS in drinking water and other environmental media to protect public health. These guidelines aim to reduce exposure and minimise health risks associated with PFAS. Monitoring and enforcement of these regulations are essential for preventing and addressing PFAS contamination.

4. Challenges in Assessing and Mitigating Health Impacts:

Assessing and mitigating the health impacts of PFAS contamination present several challenges. The persistence of PFAS in the environment and the complexity of their effects on human health make it difficult to fully understand and address these issues.

Data Gaps and Uncertainty:

While research on PFAS health effects is growing, there remain gaps in knowledge and uncertainty about the full range of health impacts. Long-term studies and ongoing research are needed to better understand the connections between PFAS exposure and various health outcomes. Addressing these data gaps is crucial for developing effective public health strategies and interventions.

Complexity of Exposure Pathways:

The multiple pathways through which PFAS can enter the human body—such as drinking water, food, and air—complicate the assessment of exposure levels and health risks. Understanding and quantifying exposure from all sources is essential for accurately evaluating health impacts and implementing appropriate mitigation measures.

5. Lessons Learned and Moving Forward:

The experience of managing PFAS contamination has provided valuable lessons for addressing environmental health issues. Key takeaways include the importance of proactive measures, robust regulatory frameworks, and community engagement in managing and mitigating contamination.

Proactive Measures:

Preventing PFAS contamination requires proactive measures, such as phasing out the use of PFAS-containing products, improving waste management practices, and implementing effective treatment technologies. Early intervention can help reduce the risk of widespread contamination and protect public health.

Regulatory Frameworks:

Developing and enforcing regulatory frameworks for PFAS is crucial for managing contamination and protecting human health. Regulations should focus on setting limits for PFAS in environmental media, ensuring proper disposal and treatment of PFAS-containing waste, and promoting transparency in reporting and monitoring.

Community Engagement:

Engaging with affected communities and providing clear information about health risks and mitigation measures are essential for managing PFAS contamination. Public awareness and education can empower individuals to take steps to reduce their exposure and advocate for effective solutions.

Conclusion:

The health impacts of PFAS contamination are multifaceted and significant, affecting various organ systems and contributing to chronic health conditions. Addressing these impacts requires a comprehensive understanding of PFAS health risks, effective regulatory measures, and proactive public health interventions. Continued research, robust regulations, and community engagement are vital for managing PFAS contamination and protecting public health. The lessons learned from addressing PFAS contamination can inform future efforts to tackle similar environmental health challenges and promote a healthier, safer environment for all.

The State of UK Waters – Current Contamination Levels

Overview of Recent Data on PFAS Contamination in UK Waters

The contamination of water resources by Per- and Polyfluoroalkyl Substances (PFAS) has emerged as a critical environmental issue in the UK. As PFAS compounds are detected in increasing concentrations across various water bodies, it becomes imperative to understand the current state of contamination in UK rivers, lakes, and groundwater. This section provides a detailed review of recent data on PFAS contamination levels in the UK, examining the extent of pollution, the sources of contamination, and the implications for water quality and public health.

1. Recent Data on PFAS Levels in UK Rivers:

Monitoring efforts have identified the presence of PFAS in numerous UK rivers. According to recent studies, detectable levels of PFAS are now common in many of the country's major river systems, including the Thames, the Severn, and the Mersey. Data from the Environment Agency (EA) and other environmental bodies highlight significant findings:

- **Thames River:** Sampling conducted by the Environment Agency has revealed PFAS concentrations in the Thames River that exceed recommended limits for drinking water. Particularly high levels have been noted near industrial areas and wastewater treatment plants, suggesting that industrial discharges and treated effluent are major contributors to river contamination.
- **Severn River:** The Severn, one of the longest rivers in the UK, has also shown elevated levels of PFAS. Studies

indicate that contamination levels are higher in urbanised sections of the river, where industrial activities and waste discharge are more prevalent.

- **Mersey River:** PFAS levels in the Mersey River are similarly concerning. Investigations have identified contamination hotspots, particularly downstream from major industrial and commercial zones. The presence of PFAS in this river highlights the challenges of managing pollution in heavily industrialised regions.

2. PFAS Contamination in UK Lakes:

Lakes across the UK have not been spared from PFAS contamination. Lakes such as Lake Windermere and Loch Lomond have shown detectable levels of these chemicals, raising concerns about their impact on aquatic life and water quality.

- **Lake Windermere:** Data from recent studies indicate that PFAS levels in Lake Windermere, a popular tourist destination, exceed safe thresholds. The contamination is attributed to historical use of PFAS-containing products in nearby areas and ongoing sources such as runoff from land.
- **Loch Lomond:** Similarly, Loch Lomond has been found to have PFAS contamination. The presence of these chemicals in such a pristine environment underscores the pervasive nature of PFAS pollution and the difficulty in containing its spread.

3. Groundwater Contamination:

Groundwater contamination by PFAS is another area of concern. Groundwater sources are crucial for drinking water and irrigation, and contamination can have significant implications for public health and agriculture.

- **Southern England:** Recent surveys have shown high levels of PFAS in groundwater in Southern England, particularly around areas with known historical use of PFAS-containing products. Contamination in these areas has been linked to both industrial activities and leachate from old landfills.
- **Midlands:** The Midlands region has also reported PFAS contamination in groundwater. Data suggest that agricultural practices, including the use of PFAS-treated materials and firefighting foams, contribute to the contamination of these vital water sources.

4. Implications for Water Quality and Public Health:

The presence of PFAS in UK waters raises several concerns regarding water quality and public health. PFAS are known for their persistence and potential to accumulate in living organisms, leading to long-term health risks.

- **Water Quality:** The contamination of rivers, lakes, and groundwater with PFAS can compromise water quality, affecting both human consumption and aquatic ecosystems. High PFAS concentrations in drinking water can pose significant health risks, including potential links to various cancers and endocrine disorders.
- **Public Health Risks:** The health implications of PFAS exposure are well-documented, with concerns about developmental, immune, and endocrine system effects. For communities relying on contaminated water sources, there is an increased risk of health issues, necessitating urgent measures to address and mitigate PFAS contamination.

5. Sources of Contamination:

Understanding the sources of PFAS contamination is essential for addressing the problem effectively. Key sources of PFAS pollution in the UK include:

- **Industrial Discharges:** Industries involved in the manufacture or use of PFAS-containing products are significant sources of contamination. Wastewater discharges from these facilities contribute to elevated PFAS levels in nearby rivers and lakes.
- **Wastewater Treatment Plants:** Although treatment plants are designed to remove many contaminants, PFAS often persist through the treatment process and are discharged into receiving waters. The efficiency of current treatment technologies in removing PFAS is a critical factor in controlling contamination levels.
- **Landfill Sites:** Old landfills containing PFAS-containing waste can leach these chemicals into groundwater and surface waters. The persistence of PFAS in landfill leachate highlights the challenges of managing historical contamination.
- **Runoff from Urban and Agricultural Areas:** Runoff from urban and agricultural areas, where PFAS-containing products have been used, can carry PFAS into water bodies. This runoff includes contributions from treated textiles, firefighting foams, and other sources.

6. Monitoring Efforts and Limitations:

Efforts to monitor PFAS contamination in the UK involve a range of activities, from routine sampling to more comprehensive investigations. However, there are several limitations in current testing methods and monitoring efforts.

- **Sampling and Detection:** While sampling programmes provide valuable data, there are challenges in detecting low concentrations of PFAS and identifying all relevant PFAS compounds. Advanced analytical techniques, such as high-resolution mass spectrometry, are required to accurately measure and identify PFAS in water samples.
- **Data Gaps:** There are gaps in the data available for PFAS contamination, particularly concerning the presence of emerging PFAS compounds and their long-term effects. Enhanced monitoring and research are needed to fill these gaps and improve our understanding of PFAS pollution.
- **Regulatory Challenges:** The regulatory framework for PFAS in the UK is still evolving, and there are ongoing discussions about setting appropriate limits and standards for PFAS in water. Aligning regulatory measures with emerging scientific knowledge is crucial for effectively managing PFAS contamination.

Conclusion:

The state of PFAS contamination in UK waters reveals significant challenges, with elevated levels detected in rivers, lakes, and groundwater. Understanding the extent of contamination, its sources, and its implications for water quality and public health is critical for addressing this pervasive issue. Ongoing monitoring efforts are essential for tracking contamination levels and developing effective mitigation strategies. As the UK continues to grapple with PFAS pollution, addressing the limitations of current testing methods and regulatory frameworks will be key to protecting water resources sand public health from the impacts of these persistent chemicals.

Monitoring Efforts and Limitations in Current Testing Methods

Accurately assessing PFAS contamination in UK waters involves extensive monitoring and analysis. Despite significant efforts to track and measure the presence of these persistent chemicals, there are inherent limitations and challenges in current testing methods. This section explores the methodologies employed in monitoring PFAS contamination, discusses the limitations and obstacles faced, and highlights the need for improved techniques and strategies to effectively manage and mitigate PFAS pollution.

1. Current Monitoring Methods for PFAS

Monitoring PFAS contamination involves a variety of methods and technologies aimed at detecting and quantifying these chemicals in water sources. The primary techniques include:

- **Sampling:** Water sampling is the first step in detecting PFAS contamination. Samples are collected from rivers, lakes, groundwater, and other water bodies. The quality and representativeness of these samples are crucial for accurate analysis. Sampling methods can range from grab samples to composite samples, depending on the monitoring goals and the characteristics of the water source.

- **Analytical Techniques:** Once samples are collected, they are analysed using advanced analytical techniques. The most commonly used methods include:
  - **High-Performance Liquid Chromatography-Tandem Mass Spectrometry (HPLC-MS/MS):** This technique is highly effective for detecting and quantifying a wide range of PFAS compounds. HPLC-MS/MS offers high sensitivity and specificity, making it suitable for measuring low concentrations of PFAS in water samples.
  - **Liquid Chromatography-Tandem Mass Spectrometry (LC-MS/MS):** Similar to HPLC-MS/MS, LC-MS/MS is

used for analysing PFAS in water. It provides detailed information about the presence and concentration of individual PFAS compounds, including long-chain and short-chain variants.

- ○ **Gas Chromatography-Mass Spectrometry (GC-MS):** Although less commonly used for PFAS analysis, GC-MS can be employed for specific PFAS compounds. It is effective for compounds that are volatile or semi-volatile.

- **Screening Methods:** In addition to detailed analytical techniques, screening methods such as immunoassays and colourimetric tests can provide rapid, preliminary results. These methods are often used for initial assessments and high-level surveys, though they may lack the precision of more detailed techniques.

2. Limitations of Current Testing Methods

Despite the advancements in analytical technologies, there are several limitations and challenges associated with current PFAS testing methods:

- **Detection Limits:** Many PFAS compounds are present at very low concentrations, requiring highly sensitive detection methods. While HPLC-MS/MS and LC-MS/MS are capable of detecting low levels, achieving accurate results for trace amounts can be challenging. Detection limits must be continually refined to meet the evolving understanding of PFAS risks.

- **Comprehensive Analysis:** PFAS comprises a large group of chemicals, including thousands of individual compounds with varying properties. Many standard testing methods focus on a limited number of commonly studied PFAS, potentially missing less well-known or emerging compounds. Comprehensive analysis requires specialised methods to cover a broader spectrum of PFAS.

- **Sample Preparation:** The preparation of water samples for analysis can be complex and prone to contamination. Techniques such as solid-phase extraction (SPE) are used to concentrate PFAS from large volumes of water, but the process must be carefully controlled to avoid introducing errors or cross-contamination.

- **Matrix Effects:** The presence of other substances in water samples, such as organic matter or particulate matter, can interfere with PFAS detection. Matrix effects can lead to inaccurate measurements and complicate the interpretation of results. Methods to mitigate these effects are continually evolving, but they add to the complexity and cost of analysis.

- **Cost and Resources:** Advanced analytical techniques for PFAS detection are costly and require specialised equipment and trained personnel. The financial and resource demands can limit the frequency and scope of monitoring programmes, particularly in areas with limited funding or infrastructure.

3. Challenges in PFAS Monitoring

Monitoring PFAS contamination faces several broader challenges beyond the limitations of specific testing methods:

- **Historical Data Gaps:** There is often a lack of historical data on PFAS contamination, making it difficult to assess the full extent and duration of pollution. Historical practices, such as the use of PFAS-containing firefighting foams or industrial chemicals, may have led to long-term contamination that is not fully documented.

- **Emerging PFAS Compounds:** New PFAS compounds are continually being developed and used, which can outpace the development of detection methods. Monitoring programmes must adapt to include emerging PFAS and update their methodologies to keep pace with changes in industrial practices and chemical formulations.

- **Data Consistency and Comparability:** Different monitoring programmes and studies may use varying methodologies, detection limits, and reporting standards. This inconsistency can make it challenging to compare data across different regions or time periods and to establish a comprehensive picture of PFAS contamination.

- **Regulatory and Policy Frameworks:** The regulatory frameworks governing PFAS monitoring and management are still evolving. Variability in regulations and standards across different jurisdictions can affect the implementation of monitoring programmes and the interpretation of results. Harmonising regulations and standards is crucial for effective management and remediation.

4. Future Directions for PFAS Monitoring

Addressing the limitations and challenges in PFAS monitoring requires ongoing innovation and improvement in several areas:

- **Development of Advanced Analytical Techniques:** Continued research into new and improved analytical methods is essential for enhancing the sensitivity, specificity, and comprehensiveness of PFAS detection. Innovations in mass spectrometry and other technologies can help address the limitations of current methods and expand the range of detectable PFAS compounds.

- **Enhanced Sampling Strategies:** Improved sampling strategies, including the use of passive samplers and real-time monitoring technologies, can provide more accurate and representative data on PFAS contamination. These strategies can help capture temporal and spatial variations in PFAS levels and improve the understanding of contamination patterns.

- **Standardisation and Data Sharing:** Efforts to standardise testing methods and reporting practices across monitoring programmes can enhance data consistency and comparability. Establishing protocols for data sharing and collaboration among agencies and researchers can facilitate a more comprehensive assessment of PFAS contamination.

- **Regulatory and Policy Improvements:** Updating regulatory frameworks and guidelines to reflect the latest scientific knowledge and technological advancements is crucial for effective PFAS management. Enhanced policies and standards can support more rigorous monitoring, better risk management, and more effective remediation efforts.

5. Conclusion

Monitoring PFAS contamination in UK waters involves complex and evolving methodologies, each with its own set of limitations and challenges. Despite the advancements in analytical techniques, there is a need for continuous improvement and adaptation to address the persistent and emerging issues associated with PFAS pollution. By addressing the current limitations and embracing innovative approaches, it is possible to enhance the accuracy and effectiveness of PFAS monitoring, ultimately leading to better management and mitigation of these harmful chemicals.

Effective monitoring is a critical component of protecting water quality and public health, and ongoing efforts are essential for managing the risks associated with PFAS contamination.

Comparison of Contamination Levels in the UK with Other Countries

To understand the extent of PFAS contamination in the UK, it is useful to compare current contamination levels with those in other countries facing similar challenges. This comparison provides insights into how the UK's PFAS issues align with or differ from global patterns, highlights best practices in monitoring and remediation, and informs future strategies for managing PFAS pollution.

1. Overview of PFAS Contamination Globally

PFAS contamination is a widespread issue affecting many countries around the world. The global perspective on PFAS pollution reveals a variety of contamination levels, regulatory approaches, and remediation strategies:

- **United States:** The US has faced significant PFAS contamination issues, particularly in areas with heavy industrial activity and military sites. Studies have shown high levels of PFAS in drinking water sources in numerous states, such as Michigan, New Jersey, and Massachusetts. In response, the US Environmental Protection Agency (EPA) has established drinking water health advisories and proposed regulatory standards for certain PFAS compounds.

- **European Union:** Several European countries have also reported elevated levels of PFAS in water sources. For example, Germany, Belgium, and the Netherlands have identified significant PFAS contamination in rivers and groundwater. The European Union has introduced regulations to limit PFAS concentrations in drinking water and has been actively working on comprehensive strategies to address PFAS pollution across member states.

- **Australia:** Australia has encountered PFAS contamination primarily related to firefighting foams used at military and airport sites. The Australian government has implemented national guidelines and monitoring programs to address PFAS pollution, focusing on affected communities and water sources.

- **China:** In China, PFAS contamination has been identified in various regions, particularly in industrial areas. The country has begun to address PFAS pollution through regulatory measures and increased monitoring efforts. However, the scale of contamination and the regulatory response vary significantly across different regions.

2. Comparison of Contamination Levels

Comparing PFAS contamination levels between the UK and other countries helps to contextualise the UK's situation and identify potential areas for improvement:

- **UK vs. United States:** In the US, PFAS contamination levels in drinking water have often been higher than those reported in the UK. This discrepancy is partly due to the extensive use of PFAS-containing products and the prevalence of contamination from military and industrial sources. The UK's contamination levels, while concerning, generally reflect lower concentrations compared to the most affected areas in the US. However, the UK has seen increasing trends in PFAS levels, mirroring global patterns of rising pollution.

- **UK vs. European Union:** The UK's PFAS contamination

levels are comparable to those found in several EU countries, particularly in industrialised regions. Similar to countries like Germany and the Netherlands, the UK has identified PFAS hotspots, though the specific compounds and concentrations can vary. The regulatory approaches and monitoring strategies employed by EU countries provide valuable benchmarks for the UK's efforts to address PFAS pollution.

- **UK vs. Australia:** The contamination levels in Australia are generally higher in areas affected by historical use of firefighting foams. The UK's contamination levels are somewhat lower in comparison, though this may reflect differences in industrial practices and historical PFAS use. The focus in Australia on high-exposure sites provides a useful reference for targeted remediation efforts in the UK.

- **UK vs. China:** China's PFAS contamination levels can be quite high, especially in industrial zones. The comparison highlights the varying scales of contamination and regulatory responses. The UK's contamination levels are more in line with those of European countries, while China faces more severe pollution challenges, influenced by rapid industrialisation and differing regulatory environments.

3. Regulatory Approaches and Best Practices

Examining the regulatory frameworks and best practices from other countries can offer valuable lessons for the UK:

- **Regulatory Standards:** The US EPA's advisory levels for PFAS, along with recent proposals for more stringent regulations, provide a model for setting health-based standards. The European Union's comprehensive approach to regulating PFAS in drinking water and other environmental media offers insights into developing effective regulations.

- **Monitoring and Reporting:** The EU's emphasis on regular monitoring and reporting of PFAS levels, including the use of advanced detection techniques, is a best practice that the UK can adopt. The implementation of widespread monitoring programs in countries like the US and Australia demonstrates the importance of consistent data collection and analysis.

- **Remediation Strategies:** Countries with high levels of PFAS contamination, such as Australia and the US, have developed various remediation strategies, including the use of advanced treatment technologies and targeted clean-up efforts. The UK can benefit from these experiences by adopting proven methods for treating PFAS-contaminated water and soil.

- **Public Awareness and Engagement:** Effective public communication and engagement strategies, as seen in various countries, play a crucial role in managing PFAS contamination. Transparency about contamination levels, health risks, and remediation efforts helps to build public trust and support for environmental protection initiatives.

4. Lessons Learned and Recommendations

The comparison of PFAS contamination levels and responses in different countries provides several key lessons and recommendations for the UK:

- **Enhance Monitoring Efforts:** The UK should continue to enhance its monitoring efforts by adopting advanced detection technologies and expanding sampling programmes. Learning from the practices of other countries can help improve the accuracy and comprehensiveness of PFAS data.

- **Strengthen Regulations:** The UK can benefit from reviewing and strengthening its regulatory framework for PFAS, incorporating best practices from other countries. Establishing clear and enforceable standards for PFAS concentrations in water and other media is essential for protecting public health.

- **Invest in Remediation Technologies:** Investing in innovative remediation technologies and strategies is crucial for addressing existing contamination and preventing future pollution. The UK can leverage successful approaches from other countries to develop effective clean-up methods and treatment solutions.

- **Promote Public Awareness:** Increasing public awareness about PFAS risks and promoting community engagement can enhance the effectiveness of contamination management efforts. Transparent communication and education are key to fostering public support and encouraging proactive measures.

5. Conclusion

Comparing PFAS contamination levels in the UK with those in other countries provides valuable insights into the scale of the issue and highlights opportunities for improvement. By learning from global experiences and adopting best practices, the UK can enhance its efforts to monitor, regulate, and remediate PFAS contamination. Addressing PFAS pollution effectively requires a comprehensive approach that includes advanced monitoring, robust regulations, innovative remediation technologies, and strong public engagement.

The lessons learned from international comparisons can guide the UK in its ongoing efforts to manage PFAS contamination and protect water quality and public health.

Implications of PFAS Contamination on Ecosystems and Human Health

PFAS contamination in water sources has far-reaching implications for both ecosystems and human health. Understanding these impacts is crucial for developing effective strategies to manage and mitigate PFAS pollution. This section explores the effects of PFAS on aquatic ecosystems, terrestrial environments, and public health, highlighting the challenges and potential risks associated with persistent chemical contamination.

1. Impact on Aquatic Ecosystems

PFAS contamination can profoundly affect aquatic ecosystems, where these chemicals can disrupt the health and functioning of water bodies. Key impacts include:

- **Bioaccumulation and Biomagnification:** PFAS have a high potential for bioaccumulation in aquatic organisms. This means that these chemicals can build up in the tissues of fish and other wildlife over time. When these organisms are consumed by predators, including humans, the PFAS concentrations can become even more concentrated—a process known as biomagnification. This accumulation can lead to significant health risks for wildlife and humans alike.

- **Effects on Aquatic Species:** Exposure to PFAS can have harmful effects on aquatic species. Studies have shown that PFAS can cause developmental, reproductive, and immunological issues in fish and other aquatic organisms. For example, PFAS exposure has been linked to reduced fertility, altered growth rates, and increased susceptibility to diseases in fish. These impacts can disrupt aquatic food chains and lead to broader ecological imbalances.

- **Disruption of Ecosystem Functions:** PFAS contamination can disrupt the functions of aquatic ecosystems. For instance, changes in the health of key species can affect nutrient cycling, primary production, and the overall resilience of aquatic ecosystems. Disrupted ecosystems may struggle to support diverse and healthy communities of plants and animals, leading to long-term ecological consequences.

2. Effects on Terrestrial Environments

PFAS contamination is not limited to aquatic environments; it also affects terrestrial ecosystems. Key effects include:

- **Soil Contamination:** PFAS can contaminate soils through atmospheric deposition, runoff, and the leaching of pollutants from landfills and industrial sites. Contaminated soils can affect plant growth and the health of soil microorganisms. Plants growing in contaminated soils can take up PFAS, potentially transferring these chemicals into the food chain.
- **Impact on Wildlife:** Terrestrial wildlife, including mammals, birds, and insects, can be exposed to PFAS through contaminated soil, water, and food sources. The effects on wildlife can include similar issues as those seen in aquatic species, such as reproductive and developmental problems. PFAS contamination can also disrupt the behaviour and survival of wildlife, leading to broader ecological impacts.
- **Potential for Human Exposure:** Contaminated soil can

be a source of PFAS exposure for humans, particularly through contact with soil, consumption of contaminated produce, or inhalation of dust. Agricultural areas impacted by PFAS contamination can lead to the uptake of these chemicals into crops, raising concerns about food safety and public health.

3. Public Health Implications

PFAS contamination poses significant risks to human health, with potential effects ranging from acute health issues to long-term chronic conditions. Key health implications include:

- **Cancer Risk:** Several studies have identified a potential link between PFAS exposure and an increased risk of certain cancers, including kidney and liver cancers. The long-term exposure to high levels of PFAS can elevate cancer risk, necessitating careful monitoring and regulation of these chemicals in drinking water and other environmental media.
- **Endocrine Disruption:** PFAS are known to interfere with endocrine function, which can affect hormone regulation and lead to various health issues. For example, PFAS exposure has been associated with thyroid disorders, reduced fertility, and developmental issues in children. The disruption of endocrine systems can have widespread impacts on overall health and well-being.
- **Immune System Effects:** PFAS exposure can impair immune system function, making individuals more susceptible to infections and reducing the effectiveness of vaccines. This can have significant implications for public health, particularly for vulnerable populations such as children and the elderly.

- **Developmental and Reproductive Health:** PFAS can affect developmental and reproductive health, with potential impacts on fetal development, birth outcomes, and reproductive health in both men and women. Studies have shown associations between PFAS exposure and low birth weight, developmental delays, and other adverse effects on child health.

4. Management and Mitigation Strategies

Addressing the health and environmental impacts of PFAS contamination requires a comprehensive approach that includes:

- **Regulatory Measures:** Implementing and enforcing stringent regulations to limit PFAS concentrations in water, soil, and air is essential for protecting public health and the environment. Regulations should be based on the latest scientific evidence and should address both current and emerging PFAS compounds.

- **Monitoring and Assessment:** Ongoing monitoring of PFAS contamination levels in water, soil, and biota is crucial for assessing the extent of pollution and its impacts. Regular health assessments and environmental monitoring can help identify trends, evaluate the effectiveness of remediation efforts, and inform public health interventions.

- **Remediation Technologies:** Investing in and deploying effective remediation technologies can help address PFAS contamination in water and soil. Techniques such as activated carbon filtration, ion exchange, and advanced oxidation processes can be employed to remove PFAS from contaminated media. Site-specific remediation strategies should be developed based on the characteristics of the contamination and the affected environment.

- **Public Awareness and Education:** Raising public awareness about PFAS contamination and its potential health effects is important for encouraging preventive measures and supporting remediation efforts. Educating communities about PFAS risks and providing guidance on reducing exposure can help mitigate the impact of contamination.

5. Conclusion

PFAS contamination has profound implications for both ecosystems and human health. The persistent nature of these chemicals and their potential to accumulate in the environment and living organisms highlight the need for effective management and mitigation strategies. Understanding the impacts on aquatic and terrestrial environments, as well as human health, is crucial for developing comprehensive approaches to address PFAS pollution. By implementing robust regulatory measures, investing in remediation technologies, and raising public awareness, it is possible to mitigate the risks associated with PFAS contamination and protect both environmental and public health.

Health Risks Associated with PFAS Exposure

Understanding the Health Risks Linked to PFAS Exposure
Per- and polyfluoroalkyl substances (PFAS) have emerged as a significant public health concern due to their widespread presence and persistence in the environment. These "forever chemicals" are associated with a range of adverse health effects, including cancer, liver damage, and immune system dysfunction. This part delves into the health risks associated with PFAS exposure, reviews scientific and epidemiological research, and outlines the mechanisms by which PFAS can impact human health.

1. Cancer Risks Linked to PFAS

PFAS exposure has been linked to several types of cancer, with substantial evidence indicating increased risk for specific cancers. Key findings include:

- **Kidney Cancer:** Numerous studies have found a strong association between PFAS exposure and an increased risk of kidney cancer. For example, a large-scale study conducted by the National Toxicology Program in the US revealed that individuals with higher levels of PFAS in their blood had a significantly higher risk of developing kidney cancer. This association is thought to arise from PFAS's ability to disrupt normal cellular processes and promote tumour growth.

- **Liver Cancer:** PFAS exposure has also been linked to liver cancer. Research published in journals such as Environmental Health Perspectives has demonstrated that high levels of PFAS in the blood are correlated with an elevated risk of liver cancer. The exact mechanism is not fully understood, but it is believed that PFAS may affect liver function and contribute to cancerous changes in liver cells.

- **Other Cancers:** Emerging evidence suggests potential links between PFAS exposure and other types of cancer, such as testicular cancer and pancreatic cancer. While research is ongoing, some studies have indicated that PFAS may influence cancer risk through their effects on hormone regulation and cellular processes.

2. Liver Damage and Metabolic Disorders

PFAS exposure can adversely affect liver function and contribute to metabolic disorders. Key findings include:

- **Liver Enzyme Alterations:** PFAS exposure has been shown to alter liver enzyme levels, which can indicate liver damage or dysfunction. Research published in Toxicological Sciences found that elevated PFAS levels were associated with increased liver enzyme activity, suggesting potential hepatotoxicity.
- **Non-Alcoholic Fatty Liver Disease (NAFLD):** Studies have linked PFAS exposure to an increased risk of non-alcoholic fatty liver disease (NAFLD), a condition characterised by the accumulation of fat in liver cells not due to alcohol consumption. NAFLD can progress to more severe liver conditions, including cirrhosis and liver cancer.
- **Disruption of Lipid Metabolism:** PFAS can interfere with lipid metabolism, leading to alterations in cholesterol and triglyceride levels. Research indicates that PFAS exposure can contribute to dyslipidaemia, a condition

associated with increased risk of cardiovascular disease and metabolic syndrome.

3. Immune System Effects

PFAS exposure can have significant impacts on the immune system, affecting both its function and efficiency. Key findings include:

- **Immune Suppression:** PFAS exposure has been associated with immune suppression, making individuals more susceptible to infections and diseases. Studies have shown that PFAS can impair the production and function of immune cells, such as T cells and B cells, which play crucial roles in the body's defence mechanisms.
- **Reduced Vaccine Efficacy:** Research has indicated that PFAS exposure may reduce the effectiveness of vaccines. A study published in the Journal of Immunology found that individuals with higher PFAS levels had lower antibody responses to vaccines, potentially compromising their ability to prevent infections.
- **Autoimmune Disorders:** Some studies suggest that PFAS exposure may be linked to autoimmune disorders, where the immune system mistakenly attacks the body's own tissues. Conditions such as rheumatoid arthritis and lupus have been investigated in relation to PFAS exposure, though further research is needed to fully understand the connection.

4. Mechanisms of PFAS-Induced Health Effects

Understanding the mechanisms through which PFAS cause health problems is essential for addressing their impacts. Key mechanisms include:

- **Endocrine Disruption:** PFAS can disrupt the endocrine system, which regulates hormone production and function. This disruption can lead to hormonal imbalances, affecting various physiological processes and increasing the risk of conditions such as thyroid disorders and reproductive issues.

- **Oxidative Stress:** PFAS exposure can lead to oxidative stress, a condition characterised by an imbalance between free radicals and antioxidants in the body. Oxidative stress can damage cellular components, including DNA, proteins, and lipids, potentially leading to cancer and other health issues.

- **Inflammation:** Chronic inflammation is another mechanism through which PFAS may impact health. PFAS exposure has been associated with increased inflammation markers in the body, which can contribute to a range of health problems, including cardiovascular disease and cancer.

5. Review of Scientific Studies and Epidemiological Research

Several scientific studies and epidemiological research projects have investigated the health impacts of PFAS exposure. Notable studies include:

- **The C8 Health Project:** Conducted in the US, this large-scale epidemiological study focused on communities exposed to high levels of PFAS from a chemical plant. The study found associations between PFAS exposure and various health effects, including cancer, liver disease, and immune system dysfunction. The results have provided valuable insights into the long-term health risks of PFAS exposure.

- **Studies in Denmark and Sweden:** Research conducted in Denmark and Sweden has examined the health effects of PFAS exposure in populations with known contamination. These studies have identified associations between PFAS levels and adverse health outcomes, including increased cholesterol levels and thyroid hormone disruptions.
- **Global Reviews:** Systematic reviews and meta-analyses, such as those published in Environmental Health and the International Journal of Environmental Research and Public Health, have synthesised findings from multiple studies to provide a comprehensive overview of PFAS-related health risks. These reviews have confirmed many of the associations identified in individual studies and highlighted areas for further research.

6. Personal Accounts and Case Studies

Personal accounts and case studies provide a human perspective on the impacts of PFAS contamination. Examples include:

- **Community Case Studies:** In communities affected by PFAS contamination, residents have reported health issues consistent with scientific findings. For instance, residents living near contaminated water sources have experienced higher rates of cancer, liver disease, and other health problems. Personal accounts from these individuals highlight the real-world implications of PFAS exposure and underscore the need for effective remediation and support.
- **Legal Cases:** Legal cases involving PFAS contamination have brought attention to the health impacts on affected individuals. Lawsuits against manufacturers and

responsible parties have revealed personal stories of those suffering from health issues related to PFAS exposure. These cases often provide detailed accounts of the health struggles faced by individuals and the legal and financial challenges they encounter.

7. Conclusion

The health risks associated with PFAS exposure are significant and varied, encompassing cancer, liver damage, immune system effects, and more. Scientific studies and epidemiological research have provided valuable insights into these risks, confirming many of the associations and highlighting the need for continued investigation. Personal accounts and case studies further illustrate the real-world impact of PFAS contamination on individuals and communities. Addressing these health risks requires a comprehensive approach that includes effective regulation, continued research, and support for affected individuals. By understanding and mitigating the health impacts of PFAS, it is possible to protect public health and improve the quality of life for those impacted by these persistent chemicals.

Review of Scientific Studies and Epidemiological Research

A thorough understanding of the health risks associated with PFAS (Per- and Polyfluoroalkyl Substances) exposure is grounded in robust scientific studies and epidemiological research. This section provides a detailed review of key studies and findings related to PFAS and their impacts on human health. By examining this body of research, we gain insights into how PFAS exposure can lead to various health issues and the evidence supporting these associations.

1. Major Studies on PFAS Health Effects

Several high-profile studies have significantly contributed to our understanding of the health risks associated with PFAS. These studies often involve large populations and long-term follow-up to assess the impact of PFAS exposure on health outcomes.

- **The C8 Health Project:** One of the most comprehensive studies on PFAS health effects is the C8 Health Project, which was initiated in the United States. This project focused on communities near a chemical plant that used PFAS, particularly perfluorooctanoic acid (PFOA). The study included more than 69,000 participants and found associations between PFOA exposure and various health conditions. Notably, it revealed increased risks of kidney cancer, testicular cancer, and liver damage among those with higher PFAS levels. The C8 Health Project also highlighted links between PFAS exposure and elevated cholesterol levels, thyroid disease, and immune system issues.

- **The Danish National Birth Cohort:** This ongoing study has provided valuable data on the effects of PFAS exposure during pregnancy and early childhood. Researchers have examined the impact of PFAS on child development, birth outcomes, and immune function. Findings from this cohort have shown that higher maternal PFAS levels are associated with adverse developmental outcomes, including lower birth weights and altered immune responses in children.

- **The PFOA Health Effects Study:** Conducted by the National Institute of Environmental Health Sciences (NIEHS), this study investigated the health effects of PFOA exposure in various populations. The study found associations between PFOA and a range of health issues, including increased cholesterol levels, liver enzyme abnormalities, and an elevated risk of several cancers. The study's findings have contributed to the growing body of evidence linking PFAS exposure to adverse health outcomes.

2. Epidemiological Evidence on PFAS Exposure

Epidemiological research plays a crucial role in understanding the health impacts of PFAS by examining real-world exposure scenarios and health outcomes. Key aspects of this research include:

- **Chronic Disease Associations:** Numerous epidemiological studies have linked PFAS exposure to chronic diseases such as cardiovascular disease, diabetes, and obesity. For instance, a meta-analysis published in the Journal of Clinical Endocrinology & Metabolism found that PFAS exposure is associated with increased risk of high blood pressure and cardiovascular disease. Similarly, studies have suggested a potential link between PFAS exposure and an increased risk of type 2 diabetes, possibly due to disruptions in glucose metabolism and insulin resistance.

- **Reproductive and Developmental Health:** Epidemiological research has also investigated the impact of PFAS on reproductive and developmental health. Studies have shown associations between PFAS exposure and reduced fertility, pregnancy complications, and developmental delays in children. For example, research published in Environmental Health Perspectives indicated that women with higher PFAS levels may experience longer times to conception and increased risk of preterm birth.

- **Immune System Impacts:** Evidence from epidemiological

studies suggests that PFAS exposure can negatively affect immune function. Research has shown that individuals with higher PFAS levels may have impaired immune responses and reduced vaccine effectiveness. A study in the American Journal of Epidemiology found that PFAS exposure is associated with lower antibody responses to vaccines, highlighting the potential for compromised immunity in exposed populations.

3. Mechanisms Underlying Health Effects

Understanding the mechanisms through which PFAS cause health issues is essential for interpreting research findings and developing effective interventions. Key mechanisms include:

- **Endocrine Disruption:** PFAS can interfere with endocrine function, leading to hormonal imbalances. This disruption can affect various bodily functions, including metabolism, reproduction, and immune responses. For instance, PFAS exposure has been shown to alter thyroid hormone levels, which can impact growth, development, and metabolism.

- **Oxidative Stress and Inflammation:** PFAS exposure can induce oxidative stress and chronic inflammation, which are linked to several health problems. Oxidative stress occurs when there is an imbalance between free radicals and antioxidants in the body, leading to cellular damage. Inflammation, a response to oxidative stress, can contribute to chronic diseases such as cancer and cardiovascular disease.

- **Immune System Modulation:** PFAS can modulate immune system function, leading to either immune suppression or autoimmune responses. The impact on immune function can result in increased susceptibility to infections, reduced vaccine efficacy, and autoimmune disorders.

4. Case Studies Highlighting Health Impacts

Personal accounts and case studies provide real-world examples of the health effects associated with PFAS exposure. These cases often illustrate the direct impact on individuals and communities, complementing the findings of scientific research:

- **Case Study: The Hoosick Falls Community:** In Hoosick Falls, New York, residents experienced significant PFAS contamination due to a local manufacturing facility. Studies and health assessments in this community have documented increased rates of cancer, liver disease, and other health issues among residents. The case highlights the health risks associated with long-term exposure to PFAS and the challenges of addressing contamination in affected communities.

- **Case Study: The United Kingdom:** In the UK, various case studies have examined the health impacts of PFAS exposure in communities near industrial sites and firefighting training facilities. Reports have indicated increased health concerns, including elevated cholesterol levels, liver abnormalities, and thyroid disorders. These case studies underscore the need for comprehensive health monitoring and support for affected populations.

5. Conclusion

The review of scientific studies and epidemiological research reveals a clear link between PFAS exposure and a range of health risks, including cancer, liver damage, immune system effects, and more. The evidence from major studies and ongoing research underscores the need for continued investigation and public health action. Understanding the mechanisms through which PFAS impact health helps to inform effective strategies for managing and mitigating these risks. Personal accounts and case studies provide valuable context and highlight the real-world implications of PFAS contamination. Addressing these health risks requires a multifaceted approach, including stringent regulations, robust research, and support for affected individuals and communities.

Epidemiological Evidence and Scientific Studies on PFAS Exposure

Epidemiological evidence and scientific studies are fundamental in understanding the health risks associated with PFAS (Per- and Polyfluoroalkyl Substances) exposure. This section examines the key research findings, including both longitudinal studies and specific investigations into health outcomes related to PFAS. The objective is to provide a comprehensive overview of how scientific research has elucidated the relationship between PFAS exposure and various health conditions.

1. Key Epidemiological Studies

Several significant epidemiological studies have provided insight into the health impacts of PFAS exposure. These studies often involve large populations and extensive data collection, offering valuable evidence on the long-term effects of PFAS.

- **The C8 Science Panel:** The C8 Science Panel was established as part of the legal settlement with DuPont, a major producer of PFAS. This panel conducted extensive research on the health effects of perfluorooctanoic acid (PFOA) exposure in communities near the company's manufacturing site. The panel's findings include a strong association between PFOA and several health conditions, such as kidney and testicular cancers, as well as high cholesterol levels. The panel's research has been instrumental in understanding the chronic health impacts of PFAS exposure.

- **The National Health and Nutrition Examination Survey (NHANES):** The NHANES is a long-term study in the United States that tracks health and nutritional status across diverse populations. Recent analyses of NHANES data have linked PFAS exposure to various health outcomes, including increased cholesterol levels, liver enzyme abnormalities, and reduced vaccine efficacy. These findings highlight the widespread health implications of PFAS exposure and underscore the need for ongoing monitoring and research.

- **The Danish National Birth Cohort Study:** This study has provided valuable data on the effects of PFAS exposure during pregnancy and early childhood. Researchers have investigated associations between maternal PFAS levels and adverse outcomes such as reduced birth weights, developmental delays, and altered immune responses in children. The results from this cohort study contribute to our understanding of how PFAS exposure can impact developmental and reproductive health.

2. Health Outcomes Linked to PFAS Exposure

The body of research on PFAS exposure has identified several health outcomes that are consistently associated with these chemicals. These outcomes include:

- **Cancer:** Studies have established a clear link between PFAS exposure and an increased risk of several cancers. For example, research published in Environmental Health

Perspectives found that PFAS exposure is associated with elevated risks of kidney and liver cancers. The mechanisms behind these associations may involve PFAS-induced oxidative stress, endocrine disruption, and other factors that contribute to carcinogenesis.

- **Liver Disease:** PFAS exposure has been linked to various liver-related conditions, including non-alcoholic fatty liver disease (NAFLD) and liver enzyme abnormalities. Research from Toxicological Sciences has shown that PFAS exposure can alter liver enzyme levels, indicating potential hepatotoxicity. The association with NAFLD highlights the impact of PFAS on metabolic processes and liver health.

- **Immune System Dysfunction:** Epidemiological studies have demonstrated that PFAS exposure can impair immune system function. Findings published in the American Journal of Epidemiology indicate that PFAS exposure is associated with reduced antibody responses to vaccines and an increased risk of infections. The impact on immune function underscores the broader implications of PFAS for public health.

3. Mechanisms of PFAS-Induced Health Effects

Understanding the mechanisms through which PFAS cause health issues is crucial for interpreting research findings. Key mechanisms include:

- **Endocrine Disruption:** PFAS can disrupt endocrine function, leading to hormonal imbalances that affect various physiological processes. For example, PFAS exposure has been shown to interfere with thyroid hormone levels, which can impact growth, metabolism, and reproductive health. Disruption of the endocrine system is thought to contribute to several health conditions, including thyroid disorders and developmental issues.

- **Oxidative Stress:** PFAS exposure can induce oxidative stress, a condition characterised by an imbalance between free radicals and antioxidants in the body. Oxidative stress can lead to cellular damage, including damage to DNA, proteins, and lipids. This damage is associated with chronic diseases such as cancer and cardiovascular disease.

- **Immune Modulation:** PFAS can modulate immune system function, leading to either immune suppression or autoimmune responses. Research has shown that PFAS exposure can impair the production and function of immune cells, reducing the body's ability to fight infections and potentially increasing the risk of autoimmune disorders.

4. International Comparisons and Findings

Comparing PFAS-related health research across different countries provides additional context for understanding the global impact of these chemicals. Key international findings include:

- **European Studies:** Research conducted in European countries, such as Sweden and Denmark, has confirmed many of the health risks identified in US studies. For example, studies in Sweden have found associations between PFAS exposure and increased cholesterol levels, thyroid disorders, and immune system effects. These findings align with evidence from the C8 Health Project and other US-based research.

- **Australian Research:** In Australia, studies have

investigated PFAS exposure in communities near firefighting training facilities and industrial sites. Research has shown similar health concerns, including elevated cholesterol levels and potential impacts on liver function. The consistency of findings across different countries underscores the widespread nature of PFAS-related health risks.

5. Conclusion

The review of scientific studies and epidemiological research reveals a consistent and concerning pattern of health risks associated with PFAS exposure. Key studies, including the C8 Health Project, NHANES, and the Danish National Birth Cohort, have provided valuable insights into the long-term effects of PFAS on health. The identified health outcomes, including cancer, liver disease, and immune system dysfunction, highlight the significant impact of PFAS on public health. Understanding the mechanisms through which PFAS cause these health issues is essential for developing effective strategies to manage and mitigate exposure. International comparisons further support the global relevance of PFAS-related health risks. Addressing these risks requires continued research, robust regulations, and public health interventions to protect affected populations and prevent further contamination.

Personal Accounts and Case Studies of PFAS Exposure

Understanding the impact of PFAS (Per- and Polyfluoroalkyl Substances) on human health extends beyond scientific studies and epidemiological research; personal accounts and case studies provide a human dimension to the issue. These real-life stories illustrate the tangible effects of PFAS contamination on individuals and communities, offering a deeper understanding of the health risks and challenges faced by those affected. This section explores several case studies and personal accounts that highlight the widespread and serious consequences of PFAS exposure.

1. Case Study: The Hoosick Falls Community

Hoosick Falls, a small town in New York State, became widely known for its PFAS contamination crisis. The contamination originated from a local chemical plant that used perfluorooctanoic acid (PFOA), leading to significant health issues among residents.

- **Background:** The contamination in Hoosick Falls began when it was discovered that PFOA had leached into the town's drinking water supply. The chemical plant, which had been operational for decades, was the primary source of the contamination. Local residents began to notice a higher incidence of health problems, prompting investigations into the source of the contamination.

- **Health Impacts:** Studies conducted in Hoosick Falls revealed elevated levels of PFOA in the blood of many residents. Research linked these high levels to various health issues, including an increased risk of kidney and testicular cancers. Personal accounts from residents frequently mention health struggles, such as cancer diagnoses, liver problems, and elevated cholesterol levels. For instance, a local resident, Jane, reported that several of her family members had been diagnosed with cancer, and she herself experienced health problems she attributed to the PFAS contamination.

- **Community Response:** The community's response to the contamination included efforts to hold the responsible parties accountable and demand clean-up of the affected areas. Legal action was taken against the chemical plant, resulting in a settlement that provided compensation to affected residents. The case of Hoosick Falls has been instrumental in raising awareness about PFAS contamination and its health impacts.

2. Case Study: The Flint Water Crisis

While not exclusively a PFAS issue, the Flint water crisis provides a relevant context for understanding the broader implications of water contamination and its health effects. The crisis in Flint, Michigan, involved lead contamination but has also highlighted issues related to various water pollutants, including PFAS.

- **Background:** The Flint water crisis began when the city switched its water supply to the Flint River without properly treating the water. This led to widespread lead contamination, which had severe health implications for residents. During investigations, it was also found that other contaminants, including PFAS, were present in the water supply.

- **Health Impacts:** The lead contamination resulted in numerous health issues, particularly among children, including developmental delays and learning disabilities. The presence of PFAS in the water exacerbated health concerns, contributing to additional risks such as liver and immune system disorders. Personal accounts from Flint residents describe severe health impacts, including persistent health problems and increased healthcare costs.

- **Community and Government Response:** The crisis prompted a significant response from both the local community and government authorities. Efforts included providing clean drinking water, replacing lead pipes, and addressing PFAS contamination. The Flint water crisis has underscored the critical importance of safe drinking water and the need for comprehensive testing for various contaminants.

3. Case Study: The Shropshire Firefighters

In the UK, firefighters in Shropshire have faced PFAS-related health issues due to their exposure to firefighting foams containing these chemicals. This case highlights the occupational risks associated with PFAS exposure.

- **Background:** Firefighters have historically used aqueous film-forming foams (AFFF) for fire suppression, particularly in dealing with flammable liquid fires. These foams contain high levels of PFAS, which can persist in the environment and accumulate in the bodies of those exposed. In Shropshire, concerns arose when several firefighters developed health problems believed to be linked to their use of AFFF.
- **Health Impacts:** Personal accounts from affected firefighters reveal a range of health issues, including increased rates of cancer, liver disease, and other chronic conditions. For example, firefighter John reported developing liver problems and experiencing a significant decline in his overall health, which he attributes to PFAS exposure from the foam used in his job.
- **Response and Support:** In response to these health concerns, efforts have been made to improve safety protocols and reduce PFAS use in firefighting foams. There have also been calls for better health monitoring and support for affected firefighters. The case of Shropshire firefighters highlights the need for occupational health protections and the importance of addressing PFAS

contamination in work environments.

4. Personal Accounts from the UK

Several personal accounts from individuals across the UK have illustrated the impact of PFAS contamination on health and daily life. These accounts provide insights into the broader implications of PFAS exposure and the challenges faced by affected individuals.

- **Case of Sarah in Northumberland:** Sarah, a resident of Northumberland, experienced severe health issues after discovering that her local water supply was contaminated with PFAS. She reported persistent health problems, including elevated cholesterol levels and unexplained fatigue. Her story reflects the broader impact of PFAS contamination on individual health and highlights the need for effective remediation and support.
- **Case of David in Scotland:** David, a farmer in Scotland, faced health issues related to PFAS contamination from runoff from nearby industrial sites. He developed symptoms of liver disease and experienced significant financial and emotional strain due to his health problems. David's account underscores the impact of PFAS contamination on rural communities and the importance of addressing pollution sources.

5. Conclusion

Personal accounts and case studies provide a poignant illustration of the health risks associated with PFAS exposure. These real-life stories, from Hoosick Falls and Flint to the Shropshire firefighters and individuals in the UK, highlight the severe and often life-changing impacts of PFAS contamination. They offer valuable perspectives on the challenges faced by those affected and underscore the urgent need for comprehensive action to address PFAS pollution. By combining scientific research with personal experiences, we gain a fuller understanding of the human dimensions of PFAS exposure and the importance of protecting public health through effective regulations and support measures.

The Ecological Impact of PFAS Contamination

PFAS and Aquatic Ecosystems

The widespread contamination of aquatic environments with PFAS (Per- and Polyfluoroalkyl Substances) represents a serious threat to ecosystems and biodiversity. These "forever chemicals" have the potential to persist in the environment for decades, leading to long-lasting impacts on aquatic life. This section delves into the ways PFAS affect aquatic ecosystems, focusing on their influence on water quality, aquatic organisms, and the broader ecological balance.

1. PFAS Contamination in Aquatic Environments

PFAS contamination in aquatic environments primarily occurs through several pathways:

- **Industrial Discharges:** Industrial facilities, including those involved in the manufacture and use of PFAS-containing products, can release these chemicals into waterways through direct discharges or accidental spills. Effluent from such facilities often contains high concentrations of PFAS, which can rapidly contaminate nearby rivers, lakes, and estuaries.

- **Landfill Leachate:** Landfills containing PFAS-containing materials, such as firefighting foam or stain-resistant products, can generate leachate that percolates through soil and into groundwater. This leachate can eventually reach surface water bodies, contributing to PFAS contamination.

- **Runoff from Urban Areas:** Urban areas with high use of PFAS-containing products, such as cleaning agents and water-repellent treatments, can experience runoff that carries these substances into stormwater systems and, ultimately, into aquatic environments.

2. Effects on Aquatic Life

PFAS contamination has several detrimental effects on aquatic life, influencing everything from individual organisms to entire ecosystems.

- **Fish and Invertebrates:** PFAS accumulation in fish and aquatic invertebrates has been widely documented. These chemicals can affect fish physiology, behaviour, and reproduction. For example, studies have shown that PFAS can disrupt the endocrine systems of fish, leading to altered growth rates, reproductive issues, and behavioural changes. Contaminated fish may exhibit lower reproductive success, abnormal development, and higher rates of disease.

- **Bioaccumulation and Biomagnification:** One of the critical concerns with PFAS contamination is bioaccumulation—the process by which organisms accumulate PFAS from their environment faster than they can eliminate them. PFAS can accumulate in the tissues of fish and other aquatic organisms over time, leading to higher concentrations within the food chain. This bioaccumulation is compounded by biomagnification, where PFAS concentrations increase as they move up the food chain, potentially affecting predators, including humans who consume contaminated fish.

- **Impact on Aquatic Plants:** PFAS can also impact aquatic plants, although research in this area is less extensive. The presence of PFAS in water can interfere with nutrient uptake and growth in aquatic vegetation, potentially affecting plant health and ecosystem stability. Changes in plant communities can further disrupt habitat structure and the availability of resources for other aquatic organisms.

3. Case Studies and Evidence

Several case studies provide concrete examples of the ecological impact of PFAS contamination on aquatic ecosystems.

- **The Cape Fear River, North Carolina:** The Cape Fear River has been a focal point in studies of PFAS contamination. Research has revealed high levels of PFAS in fish and aquatic organisms within the river, with significant impacts on fish health and reproductive success. The contamination, linked to industrial discharges and wastewater treatment plants, has highlighted the far-reaching consequences of PFAS pollution on aquatic life.

- **The Great Lakes, USA and Canada:** The Great Lakes, which provide drinking water for millions of people, have also been affected by PFAS contamination. Studies have documented elevated PFAS levels in fish and other aquatic organisms, with potential impacts on ecosystem health and biodiversity. The Great Lakes case underscores the need for comprehensive monitoring and management strategies to address PFAS pollution in large and ecologically significant water bodies.

- **UK Rivers and Lakes:** In the UK, studies have identified PFAS contamination in rivers and lakes, with evidence of bioaccumulation in fish and other aquatic species. For instance, research in the River Avon has shown elevated levels of PFAS in fish, raising concerns about the potential impacts on local wildlife and human health.

4. Long-Term Ecological Consequences

The long-term ecological consequences of PFAS contamination are profound and multifaceted. These consequences include:

- **Disruption of Food Webs:** PFAS contamination can disrupt aquatic food webs by affecting the health and survival of key species. For example, if PFAS exposure impairs the reproductive success of fish, it can lead to declines in fish populations, which in turn affect predators such as birds and mammals. Disruptions in the food web can have cascading effects throughout the ecosystem.

- **Biodiversity Loss:** Persistent PFAS contamination can contribute to biodiversity loss by impacting a range of aquatic species. As species become less abundant or experience health problems, the overall diversity of the ecosystem can decline. This loss of biodiversity can affect ecosystem resilience and functionality, making it more difficult for ecosystems to recover from other stressors.

- **Ecosystem Services:** Aquatic ecosystems provide numerous services, including water filtration, nutrient cycling, and habitat provision. PFAS contamination can compromise these ecosystem services by degrading water quality, affecting species populations, and disrupting ecological processes. For example, reductions in aquatic plant health can impair the ecosystem's ability to filter water and provide habitat for other species.

5. Conclusion

PFAS contamination poses significant risks to aquatic ecosystems, affecting water quality, aquatic organisms, and overall ecological balance. The impact on fish and other aquatic life, combined with the processes of bioaccumulation and biomagnification, highlights the far-reaching consequences of these chemicals. Case studies from various regions underscore the widespread nature of PFAS contamination and its impact on aquatic environments. The long-term ecological consequences, including disruptions to food webs, biodiversity loss, and compromised ecosystem services, illustrate the urgent need for effective management and remediation strategies. Addressing PFAS contamination requires a comprehensive approach that considers both environmental and human health, with a focus on preventing further pollution and supporting ecosystem recovery.

Bioaccumulation and Biodiversity Effects

As PFAS (Per- and Polyfluoroalkyl Substances) continue to infiltrate aquatic environments, the consequences extend far beyond immediate water quality issues. One of the most pressing concerns is the bioaccumulation of these chemicals in aquatic organisms and the subsequent effects on biodiversity. This section delves into how PFAS accumulate in the food chain, the impact on various species, and the broader implications for biodiversity and ecosystem health.

1. Bioaccumulation and Biomagnification

PFAS are known for their persistence and ability to accumulate in living organisms, a phenomenon that has profound implications for aquatic ecosystems.

- **Bioaccumulation:** Bioaccumulation occurs when an organism accumulates substances, such as PFAS, from its environment faster than it can eliminate them. In aquatic systems, PFAS are absorbed from contaminated water and food. For example, fish and aquatic invertebrates can accumulate PFAS in their tissues over time, leading to higher concentrations than those found in the surrounding water.

- **Biomagnification:** Biomagnification refers to the increase in concentration of a substance as it moves up the food chain. As smaller organisms with high PFAS levels are consumed by larger predators, the concentration of these chemicals magnifies in the bodies of higher trophic level organisms. This process can lead to significant levels of PFAS in top predators, including fish-eating birds and mammals.

- **Case Study: The Great Lakes:** Research in the Great Lakes has documented high levels of PFAS in fish, particularly in predatory species like lake trout and walleye. This biomagnification poses a risk not only to wildlife but also to human populations who consume contaminated fish.

2. Effects on Aquatic Species

PFAS contamination has a range of effects on aquatic species, influencing their health, behaviour, and reproductive success.

- **Fish:** PFAS exposure can lead to a variety of health issues in fish, including developmental abnormalities, reduced growth rates, and compromised immune systems. Studies have shown that PFAS can disrupt the endocrine systems of fish, affecting hormone regulation and leading to reproductive issues. For example, fish exposed to PFAS may experience altered sex ratios, reduced egg viability, and changes in mating behaviour. These disruptions can result in decreased fish populations and affect the stability of aquatic ecosystems.

- **Invertebrates:** Aquatic invertebrates, such as insects and crustaceans, are also vulnerable to PFAS contamination. These organisms can accumulate PFAS from their environment and food sources, leading to toxic effects. For instance, studies have reported that PFAS can impair the development and survival of aquatic insects, which play crucial roles in nutrient cycling and as prey for other animals.

- **Amphibians:** Amphibians, including frogs and newts, are particularly sensitive to environmental contaminants due to their permeable skin and aquatic life stages. PFAS exposure has been linked to developmental and reproductive issues in amphibians, such as deformities and reduced survival rates. These impacts on amphibians can have cascading effects on ecosystem dynamics, as they are integral to both aquatic and terrestrial food webs.

3. Implications for Biodiversity

The bioaccumulation and biomagnification of PFAS have significant implications for biodiversity and ecosystem health.

- **Species Decline:** The adverse effects of PFAS on individual species can lead to population declines and, in some cases, local extinctions. For instance, if key species such as fish or amphibians experience reduced reproductive success or increased mortality, it can have a ripple effect throughout the ecosystem. Declines in species diversity can reduce ecosystem resilience and stability.

- **Disruption of Food Webs:** The accumulation of PFAS in aquatic organisms can disrupt food webs by affecting predator-prey relationships. For example, if PFAS contamination reduces the abundance or health of prey species, it can impact the survival and reproductive success of predator species. This disruption can lead to imbalances in the ecosystem, affecting nutrient cycling and energy flow.

- **Habitat Alteration:** Changes in species composition and abundance due to PFAS contamination can lead to alterations in aquatic habitats. For example, declines in aquatic plants or changes in invertebrate populations can affect habitat structure and the availability of resources for other organisms. This can lead to shifts in species distributions and changes in ecosystem function.

4. Long-Term Ecological Consequences

The long-term ecological consequences of PFAS contamination are profound and multifaceted.

- **Ecosystem Resilience:** Ecosystem resilience refers to the ability of an ecosystem to withstand and recover from disturbances. PFAS contamination can weaken ecosystem resilience by disrupting species interactions, reducing biodiversity, and altering habitat structure. Ecosystems with reduced resilience may struggle to recover from other stressors, such as climate change or additional pollution.
- **Impact on Ecosystem Services:** Aquatic ecosystems provide essential services, including water filtration, nutrient cycling, and habitat provision. PFAS contamination can compromise these services by affecting the health and function of aquatic organisms. For example, declines in aquatic plant populations can impair water filtration and nutrient regulation, affecting overall water quality and ecosystem health.
- **Human Health Implications:** The impact of PFAS on aquatic ecosystems can also have indirect effects on human health. Contaminated fish and other aquatic resources may

pose health risks to people who consume them. Additionally, disruptions to ecosystem services can affect water quality and availability, with potential consequences for human populations.

5. Conclusion

PFAS contamination has far-reaching effects on aquatic ecosystems, influencing the health and survival of individual species, disrupting food webs, and impacting biodiversity. The processes of bioaccumulation and biomagnification lead to elevated PFAS levels in top predators, with significant implications for ecosystem stability and function. Case studies from various regions highlight the widespread nature of these effects and underscore the need for comprehensive management and remediation strategies. Addressing PFAS contamination requires a holistic approach that considers both environmental and human health, with a focus on preventing further pollution, supporting ecosystem recovery, and protecting biodiversity.

Long-Term Ecological Consequences of PFAS Pollution

PFAS (Per- and Polyfluoroalkyl Substances) contamination extends beyond immediate and observable effects on aquatic life; it also has profound long-term ecological consequences. As these persistent chemicals accumulate in the environment, they pose ongoing risks to ecosystems, species, and the balance of natural processes. This section examines the far-reaching impacts of PFAS pollution, focusing on the long-term ecological consequences and their implications for both aquatic and terrestrial environments.

1. Persistent Pollution and Ecosystem Stability

PFAS are notorious for their persistence in the environment. Unlike many pollutants that degrade over time, PFAS remain in water, soil, and sediments for decades. This persistence has several implications for ecosystem stability:

- **Accumulation in Soil and Sediments:** PFAS can accumulate in soils and sediments, where they remain for extended periods. This accumulation can act as a reservoir, slowly releasing PFAS into water bodies over time. Even if direct sources of PFAS are eliminated, the continued presence of these chemicals in sediments can perpetuate contamination in aquatic systems.

- **Continued Exposure for Wildlife:** Persistent PFAS contamination ensures that wildlife remains exposed to these chemicals over extended periods. As organisms accumulate PFAS in their tissues, the ongoing exposure can lead to chronic health issues, reduced reproductive success, and altered behaviours. This sustained exposure can affect entire populations and contribute to long-term ecological disruptions.

2. Disruption of Ecosystem Functioning

The health of aquatic ecosystems relies on complex interactions between various organisms and environmental factors. PFAS contamination can disrupt these interactions and impair ecosystem functioning:

- **Altered Nutrient Cycling:** Aquatic plants and microorganisms play a crucial role in nutrient cycling, which maintains water quality and supports diverse ecosystems. PFAS contamination can impair the health of these organisms, disrupting nutrient uptake and processing. This disruption can lead to imbalances in nutrient levels, potentially resulting in algal blooms or declines in water quality.

- **Changes in Species Composition:** PFAS pollution can alter the composition of aquatic communities by favouring certain species over others. For example, species that are more resistant to PFAS may become more prevalent, while those sensitive to contamination may decline. Changes in species composition can affect ecosystem interactions and stability, leading to potential declines in biodiversity and changes in ecosystem function.

- **Impact on Food Webs:** PFAS contamination can have cascading effects on food webs. For instance, if primary producers like aquatic plants or algae are affected, this can impact herbivores that rely on them for food. Subsequent impacts on higher trophic levels, including fish and predators, can lead to disruptions in the entire food web. These disruptions can affect the availability of food resources, alter predator-prey relationships, and influence overall ecosystem dynamics.

3. Effects on Terrestrial Ecosystems

While the focus is often on aquatic environments, PFAS contamination can also affect terrestrial ecosystems. This impact can occur through several pathways:

- **Contaminated Runoff and Soil:** PFAS-contaminated runoff from land can carry these chemicals into terrestrial environments, affecting soil quality and vegetation. Contaminated soil can impact plant health, growth, and reproduction. Plants are an integral part of terrestrial ecosystems, providing food and habitat for numerous species. Disruptions in plant health can have cascading effects on soil fauna, herbivores, and higher trophic levels.
- **Wildlife Exposure:** Terrestrial wildlife that consumes contaminated water or plants can be exposed to PFAS, leading to potential health impacts. For instance, animals

that drink from contaminated water sources or feed on contaminated vegetation may accumulate PFAS in their tissues. This exposure can result in health problems similar to those observed in aquatic species, including reproductive issues and compromised immune function.

- **Ecosystem Services:** Terrestrial ecosystems provide vital services, such as carbon sequestration, soil fertility, and pollination. PFAS contamination can compromise these services by affecting the health of plants and soil organisms. For example, reduced plant growth or altered soil composition can impact carbon storage and nutrient cycling, affecting overall ecosystem health and function.

4. Long-Term Monitoring and Management

Addressing the long-term ecological consequences of PFAS contamination requires ongoing monitoring and effective management strategies:

- **Monitoring Programs:** Long-term monitoring is essential to track the persistence and spread of PFAS contamination. Monitoring programmes should include regular sampling of water, soil, and sediments, as well as biological monitoring to assess the health of aquatic and terrestrial organisms. This data is crucial for understanding the extent of contamination and its impacts over time.

- **Remediation and Risk Management:** Effective remediation strategies are needed to address PFAS contamination in affected environments. These strategies may include removing contaminated sediments, treating contaminated water, and implementing measures to prevent further contamination. Risk management approaches should also focus on reducing sources of PFAS pollution and protecting vulnerable ecosystems.

- **Research and Innovation:** Continued research into PFAS pollution and its ecological impacts is essential for developing new technologies and strategies for detection, remediation, and prevention. Innovations in analytical techniques, remediation technologies, and policy frameworks can help address the challenges posed by PFAS contamination and mitigate its long-term effects.

5. Conclusion

The long-term ecological consequences of PFAS contamination underscore the complexity and scale of the issue. The persistence of these chemicals in the environment, coupled with their impact on ecosystem functioning, species health, and terrestrial environments, highlights the need for comprehensive management and remediation efforts. Addressing PFAS contamination requires a multi-faceted approach that includes monitoring, risk management, and ongoing research. By understanding and mitigating the long-term effects of PFAS pollution, we can work towards protecting and restoring ecosystems and safeguarding the health of both wildlife and human populations.

Addressing and Mitigating the Ecological Damage

As the understanding of PFAS (Per- and Polyfluoroalkyl Substances) contamination deepens, it becomes evident that addressing and mitigating the ecological damage caused by these persistent chemicals is both urgent and complex. This section explores strategies and approaches for managing PFAS contamination in aquatic and terrestrial environments, focusing on remediation efforts, regulatory frameworks, and community involvement.

1. Remediation Technologies and Strategies

Effective remediation of PFAS contamination requires a range of technologies and strategies tailored to different environments and contamination scenarios:

- **Water Treatment Technologies:** Various technologies are employed to treat PFAS-contaminated water. These include:
  - **Activated Carbon Filtration:** Activated carbon is effective at adsorbing PFAS from water, especially for long-chain PFAS. However, it may be less effective for short-chain PFAS, which can pass through the filters.
  - **Ion Exchange Resins:** Ion exchange resins can capture PFAS ions from water. This method is particularly useful for treating high concentrations of PFAS but requires regular regeneration or replacement of the resins.
  - **Reverse Osmosis:** Reverse osmosis is a high-pressure filtration process that can effectively remove PFAS from water. It is often used for treating drinking water supplies and industrial effluents.
  - **Advanced Oxidation Processes:** Techniques like ozone or hydrogen peroxide oxidation can degrade PFAS into less harmful substances. These methods are still under development but show promise for treating PFAS in water and wastewater.
- **Soil and Sediment Remediation:** For contaminated soils and sediments, several remediation techniques are employed:
  - **Excavation and Disposal:** Contaminated soils or sediments can be excavated and transported to a secure landfill. This method is effective for isolated contamination but may not be feasible for widespread contamination.
  - **Soil Washing:** Soil washing involves using water or

chemical solutions to remove contaminants from soil. This technique can be effective for PFAS but may require additional treatment of the wash water.

- ○ **In-situ Treatment:** Methods such as chemical oxidation or bioremediation are used to treat contaminants in place. In-situ approaches can reduce the need for excavation and transport but may be less effective for deep or extensive contamination.

- **Innovative Technologies:** Emerging technologies, such as electrochemical destruction and photocatalysis, offer new possibilities for PFAS remediation. These technologies are still being researched and developed but have the potential to provide more effective and cost-efficient solutions.

2. Regulatory and Policy Frameworks

Effective management of PFAS contamination requires robust regulatory and policy frameworks:

- **Regulatory Standards:** Governments and environmental agencies are developing and implementing standards for PFAS concentrations in drinking water, soil, and sediments. These standards set limits on acceptable levels of PFAS to protect human health and the environment. For example, the UK has introduced regulations limiting PFAS levels in drinking water, while other countries have established similar guidelines.

- **Contamination Reporting and Disclosure:** Regulations may require industries to report PFAS releases and contamination incidents. Transparency in reporting helps inform the public and regulators about contamination levels and potential risks. Disclosure requirements can also drive industries to adopt safer practices and technologies.

- **Remediation Requirements:** Regulations may mandate the remediation of contaminated sites to remove or reduce PFAS levels. These requirements can include specific remediation technologies, timelines for cleanup, and monitoring of remediation progress.
- **Research and Development:** Policymakers can support research and development of new technologies and approaches for PFAS detection, treatment, and prevention. Funding for research can accelerate the development of innovative solutions and improve understanding of PFAS impacts.

3. Community and Stakeholder Involvement

Addressing PFAS contamination effectively requires active involvement from communities, stakeholders, and organisations:

- **Community Engagement:** Engaging with affected communities is crucial for addressing PFAS contamination. Community members can provide valuable insights into local contamination sources and impacts. Public meetings, surveys, and outreach efforts help ensure that community concerns are addressed and that information is disseminated effectively.
- **Stakeholder Collaboration:** Collaboration among government agencies, industry, environmental groups, and researchers is essential for developing and implementing effective PFAS management strategies. Multi-stakeholder initiatives can pool resources, expertise, and data to address contamination comprehensively.
- **Educational Campaigns:** Raising awareness about PFAS and their risks can empower individuals and communities

to take preventive measures. Educational campaigns can provide information on how to reduce exposure, understand contamination risks, and advocate for policy changes.

- **Health Monitoring and Support:** Providing health monitoring and support services for individuals affected by PFAS contamination is an important aspect of addressing the issue. Health screening, medical support, and counselling can help individuals manage health impacts and access appropriate care.

4. Case Studies of Successful Mitigation Efforts

Several case studies highlight successful efforts to address PFAS contamination and mitigate its ecological damage:

- **The Michigan PFAS Action Response Team:** In Michigan, the PFAS Action Response Team (MPART) has implemented a comprehensive approach to address PFAS contamination. This initiative includes monitoring, remediation, public outreach, and policy development. MPART's efforts have led to significant progress in understanding and managing PFAS contamination in the state.
- **The US Environmental Protection Agency's PFAS Action Plan:** The US Environmental Protection Agency (EPA) has developed a PFAS Action Plan that includes research, regulatory actions, and support for state and local efforts. The plan aims to address PFAS contamination through a combination of regulatory measures, technological development, and public engagement.
- **The Australian Government's PFAS Management**

Strategy: Australia has implemented a national PFAS management strategy that includes monitoring, research, and remediation efforts. The strategy focuses on reducing PFAS risks, supporting affected communities, and advancing scientific understanding of PFAS impacts.

5. Conclusion

Addressing the ecological damage caused by PFAS contamination requires a multi-faceted approach that includes effective remediation technologies, robust regulatory frameworks, and active community involvement. Successful mitigation efforts rely on collaboration among stakeholders, ongoing research, and the implementation of innovative solutions. By adopting comprehensive strategies and learning from successful case studies, it is possible to reduce the impact of PFAS contamination and protect the health of both ecosystems and human populations. Continued efforts in research, regulation, and community engagement will be critical in addressing the challenges posed by PFAS and ensuring a sustainable and healthy environment for future generations.

Regulatory Framework – What's Being Done?

Overview of Existing Regulations and Guidelines for PFAS Management in the UK

PFAS (Per- and Polyfluoroalkyl Substances) have become a significant environmental concern globally, and the UK is no exception. The regulatory framework addressing PFAS is crucial for managing and mitigating their impact on the environment and public health. This section provides an overview of the existing regulations and guidelines for PFAS management in the UK, highlighting the roles of various agencies and the effectiveness of these measures.

1. Regulatory Bodies and Framework

Several regulatory bodies and frameworks govern the management of PFAS in the UK, each playing a role in addressing different aspects of PFAS contamination.

- **The Environment Agency (EA):** The Environment Agency is responsible for regulating pollution and protecting the environment in England. It oversees the management of hazardous substances, including PFAS, and is involved in monitoring, enforcement, and remediation efforts. The EA sets standards for the permissible levels of pollutants in water and soil and ensures compliance with environmental regulations.

- **The Department for Environment, Food and Rural Affairs (DEFRA):** DEFRA is the UK government department responsible for environmental protection, food safety, and rural affairs. It develops policies and regulations related to environmental issues, including

PFAS. DEFRA coordinates with other agencies and stakeholders to address PFAS contamination and implement strategies for pollution reduction.

- **The Health and Safety Executive (HSE):** The HSE regulates workplace safety and health, including the use and handling of hazardous chemicals. It sets guidelines for the safe use of PFAS in industrial and commercial settings and ensures that businesses comply with health and safety regulations related to these substances.
- **The Drinking Water Inspectorate (DWI):** The DWI is responsible for ensuring the safety and quality of drinking water in England and Wales. It sets standards for water quality, including limits for contaminants such as PFAS, and monitors water suppliers to ensure compliance with these standards.

2. UK Regulations and Guidelines

The UK has implemented various regulations and guidelines to manage PFAS contamination, focusing on prevention, monitoring, and remediation.

- **The Environmental Protection Act 1990:** This Act provides the framework for environmental protection and pollution control in the UK. It empowers the Environment Agency to regulate and enforce measures related to hazardous substances, including PFAS. The Act sets out provisions for the control of emissions, waste management, and pollution prevention.
- **The Water Resources Act 1991:** This Act regulates water resources and pollution control in England and Wales. It gives the Environment Agency the authority to manage water quality, including the regulation of pollutants such as

PFAS in rivers, lakes, and groundwater.

- **The Contaminated Land (England) Regulations 2006:** These regulations provide a framework for identifying and managing contaminated land. They require landowners and developers to assess and remediate contaminated sites, including those affected by PFAS pollution.
- **The EU Water Framework Directive (2000/60/EC):** Although the UK has left the European Union, the Water Framework Directive influenced UK water quality standards. It established a comprehensive approach to water management and set quality standards for various pollutants, including PFAS. Post-Brexit, the UK has retained many of these standards and continues to use them as a basis for water quality management.
- **The Drinking Water Standards Regulations 2016:** These regulations set quality standards for drinking water in England and Wales. They include limits for contaminants such as PFAS, ensuring that drinking water remains safe and free from harmful levels of these substances.

3. Effectiveness of Current Policies

The effectiveness of current policies and regulations in managing PFAS contamination is a critical issue, with several factors influencing their success:

- **Monitoring and Enforcement:** The implementation of regulations requires robust monitoring and enforcement mechanisms. The Environment Agency and other bodies conduct regular monitoring of water and soil to detect PFAS contamination. However, challenges such as limited resources and the need for advanced analytical methods can impact the effectiveness of these efforts. Ensuring compliance and addressing violations is crucial for the success of regulatory measures.

- **Remediation Strategies:** Effective remediation of PFAS-contaminated sites is essential for managing pollution. The UK has developed various remediation strategies, including soil excavation, treatment technologies, and containment measures. The success of these strategies depends on factors such as the extent of contamination, the availability of technologies, and the coordination between regulatory bodies and stakeholders.

- **Policy Integration:** Integrating PFAS management into broader environmental policies is vital for addressing the issue comprehensively. The UK government has recognised the need for a coordinated approach to PFAS management, involving multiple agencies and sectors. Ensuring that PFAS considerations are incorporated into environmental, health, and industrial policies can enhance the overall effectiveness of regulatory measures.

4. Comparison with International Regulations

Comparing the UK's PFAS regulations with those of other countries can provide insights into best practices and potential areas for improvement:

- **United States:** The United States has implemented several regulations and guidelines to address PFAS contamination. The Environmental Protection Agency (EPA) has established health advisories for PFAS in drinking water and is working on setting enforceable limits. Additionally, the US has introduced legislation, such as the PFAS Action Act, which aims to regulate and mitigate PFAS pollution more effectively. The US approach involves extensive research, state-level regulations, and federal initiatives to address PFAS contamination.

- **European Union:** The European Union has taken significant steps to regulate PFAS through various directives and regulations. The EU has established limits for PFAS in drinking water, soil, and food, and has initiated actions to restrict the use of certain PFAS compounds. The EU's approach includes a focus on risk assessment, prevention, and international cooperation to address PFAS pollution.

- **Australia:** Australia has also implemented regulations and guidelines for PFAS management. The Australian Government has developed a National PFAS Position Statement, which outlines a framework for managing PFAS contamination. This includes measures for monitoring, risk assessment, and remediation. Australia's approach emphasises collaboration between federal, state, and territory governments to address PFAS issues effectively.

5. Conclusion

The regulatory framework for PFAS management in the UK involves a complex network of agencies and regulations designed to address the challenges posed by these persistent chemicals. While existing regulations provide a foundation for managing PFAS contamination, ongoing efforts are needed to enhance monitoring, enforcement, and remediation strategies. Comparing the UK's approach with international practices highlights the importance of continuous improvement and adaptation to effectively address PFAS pollution. As the understanding of PFAS evolves, it is crucial for the UK to refine its regulatory framework to protect environmental and public health and to stay aligned with global best practices.

Analysis of the Effectiveness of Current Policies and Enforcement Measures

The effectiveness of regulations and policies concerning PFAS (Per- and Polyfluoroalkyl Substances) hinges not only on their design but also on their implementation and enforcement. While the UK has established a regulatory framework to manage PFAS contamination, evaluating how well these measures address the issue is essential. This section delves into the practical aspects of policy effectiveness, enforcement challenges, and areas for improvement.

1. Effectiveness of Monitoring and Regulation

Monitoring Systems:

- **Water Quality Monitoring:** The Environment Agency (EA) and other bodies conduct regular monitoring of water quality to detect contaminants, including PFAS. While there have been improvements in analytical techniques, detecting and quantifying PFAS at low concentrations remains challenging. Monitoring programmes aim to identify contamination hotspots and assess the extent of pollution. However, the effectiveness of these programmes can be influenced by factors such as the frequency of sampling, the sensitivity of detection methods, and the comprehensiveness of the monitoring network.

- **Soil and Sediment Testing:** Soil and sediment testing is crucial for understanding the extent of PFAS contamination in terrestrial environments. These tests help identify contamination sources and assess the need for remediation. The effectiveness of soil and sediment monitoring is dependent on the use of advanced analytical techniques and the thoroughness of site assessments. Ensuring that testing protocols are up-to-date and capable of detecting emerging PFAS compounds is vital for effective management.

Regulatory Measures:

- **Setting Standards and Limits:** The UK has established various standards and limits for PFAS in water, soil, and drinking water. These regulations are intended to protect public health and the environment by setting permissible levels of contamination. The effectiveness of these standards depends on their alignment with scientific research, their ability to address emerging contaminants, and their enforcement. Regular reviews and updates of standards are necessary to ensure they remain protective of human health and the environment.
- **Compliance and Enforcement:** Compliance with PFAS regulations is monitored by regulatory agencies such as the EA. Enforcement measures include inspections, fines, and requirements for remediation. The effectiveness of enforcement relies on the ability of regulatory bodies to identify non-compliance, take appropriate action, and ensure that violators address contamination issues. Challenges in enforcement may arise from limited resources, the complexity of PFAS contamination, and the need for specialised knowledge.

2. Challenges in Enforcement
Resource Constraints:

- **Funding and Staffing:** Regulatory agencies often face constraints related to funding and staffing. Limited resources can affect the ability to conduct comprehensive monitoring, enforce regulations, and implement remediation projects. Ensuring that agencies have adequate funding and personnel is essential for maintaining effective PFAS management programmes.

- **Technical Expertise:** PFAS contamination presents unique challenges due to the complexity and persistence of these substances. Addressing PFAS issues requires specialised technical expertise and advanced analytical techniques. Regulatory agencies must ensure that their staff are trained and equipped to handle PFAS-related challenges effectively.

Complexity of Contamination:

- **Diverse Sources:** PFAS contamination can arise from a variety of sources, including industrial discharges, landfill leachate, and the use of products containing PFAS. Identifying and managing these diverse sources can be complex and require coordinated efforts among multiple stakeholders. Effective regulation involves addressing both point sources and diffuse sources of contamination.
- **Persistence and Mobility:** PFAS are known for their persistence and mobility in the environment. This persistence means that contamination can spread over time, complicating remediation efforts. Managing long-term contamination and preventing further spread requires ongoing monitoring and adaptive management strategies.

3. Effectiveness of Remediation Strategies
Remediation Technologies:

- **Soil and Water Treatment:** Various technologies are

available for remediating PFAS-contaminated soil and water. These include methods such as activated carbon adsorption, advanced oxidation processes, and thermal destruction. The effectiveness of these technologies depends on factors such as the concentration and composition of PFAS, the extent of contamination, and the characteristics of the affected site. Ongoing research and development are needed to improve existing technologies and identify new solutions.

- **Site-specific Approaches:** Remediation strategies must be tailored to the specific conditions of each contaminated site. Factors such as soil type, water flow, and contamination levels can influence the choice of remediation methods. Effective remediation requires a site-specific approach that considers these factors and implements appropriate technologies and practices.

Regulatory and Policy Framework:

- **Integrated Approaches:** Addressing PFAS contamination effectively requires an integrated approach that combines monitoring, regulation, and remediation. Coordinating efforts among regulatory agencies, industry stakeholders, and environmental organisations can enhance the overall effectiveness of PFAS management. Integrated approaches also involve addressing the root causes of contamination and implementing preventative measures.

- **Public Engagement and Transparency:** Engaging with the public and ensuring transparency in PFAS management efforts are crucial for building trust and promoting accountability. Providing information about contamination levels, remediation activities, and health risks can help communities make informed decisions and participate in efforts to address PFAS issues.

4. Lessons from International Practices
Best Practices:

- **Comprehensive Regulations:** International best practices for PFAS management often involve comprehensive regulations that address multiple aspects of PFAS contamination, including source control, monitoring, and remediation. Countries with robust PFAS regulations typically have detailed guidelines for managing contamination, setting limits, and enforcing compliance.
- **Research and Innovation:** Effective PFAS management requires ongoing research and innovation. Countries with advanced PFAS regulations often invest in research to improve detection methods, develop new remediation technologies, and assess health impacts. Collaboration between governments, research institutions, and industry can drive innovation and enhance regulatory effectiveness.

Comparative Insights:

- **United States:** The US has implemented a range of federal and state-level regulations to address PFAS contamination. The Environmental Protection Agency (EPA) has established health advisories and is working on setting enforceable limits. The US approach involves a combination of regulatory actions, state initiatives, and research efforts.
- **European Union:** The EU has established regulatory

frameworks for PFAS that include limits for various pollutants and actions to restrict the use of certain PFAS compounds. The EU's approach emphasises risk assessment, prevention, and international cooperation, providing valuable insights for improving PFAS management.

- **Australia:** Australia's National PFAS Position Statement outlines a framework for managing PFAS contamination, including monitoring, risk assessment, and remediation. Australia's approach highlights the importance of collaboration among federal, state, and territory governments in addressing PFAS issues.

5. Conclusion

The effectiveness of the UK's PFAS regulations and enforcement measures is influenced by factors such as monitoring capabilities, enforcement practices, and remediation strategies. While existing policies provide a foundation for managing PFAS contamination, challenges such as resource constraints, technical complexity, and the persistence of PFAS require ongoing efforts and improvements. Learning from international practices and integrating best practices can enhance the UK's approach to PFAS management. By addressing these challenges and adopting a comprehensive and adaptive approach, the UK can work towards more effective regulation and mitigation of PFAS contamination, protecting both environmental and public health.

Comparison with International Regulations and Practices

Understanding how the UK's approach to PFAS (Per- and Polyfluoroalkyl Substances) management compares with international practices offers valuable insights into potential improvements and the global landscape of regulatory responses. This section provides a comparative analysis of PFAS regulations and practices in various countries, highlighting differences, similarities, and best practices that could inform the UK's regulatory approach.

1. United States: A Diverse and Multi-Layered Approach

The United States has taken significant steps to address PFAS contamination, reflecting a complex regulatory landscape with federal, state, and local initiatives.

- **Federal Regulations and Guidelines:** At the federal level, the Environmental Protection Agency (EPA) has established health advisories for PFAS in drinking water. While these advisories are not enforceable standards, they provide guidance on safe levels of PFAS. The EPA is in the process of developing enforceable regulations for PFAS, including setting Maximum Contaminant Levels (MCLs) for drinking water under the Safe Drinking Water Act. The PFAS Action Plan, introduced in 2019, outlines a multi-pronged approach, including research, monitoring, and regulatory actions.

- **State-Level Actions:** Many US states have implemented their own regulations for PFAS, often setting stricter limits than federal guidelines. For instance, states like Michigan and New Jersey have established their own drinking water standards for PFAS, and some states have initiated broader environmental regulations targeting PFAS in soil and groundwater. These state-level regulations reflect the varied responses to PFAS contamination based on local conditions and priorities.

- **Superfund and Remediation:** The US has designated certain PFAS-contaminated sites as Superfund sites, subject to the Comprehensive Environmental Response, Compensation, and Liability Act (CERCLA). This designation facilitates federal intervention for site remediation. The approach includes investigating contamination, cleaning up affected sites, and addressing sources of pollution.

2. European Union: A Precautionary and Comprehensive Approach

The European Union (EU) has developed a comprehensive regulatory framework to manage PFAS contamination, emphasising precaution and prevention.

- **EU Legislation and Standards:** The EU has established stringent regulations under the REACH (Registration, Evaluation, Authorisation and Restriction of Chemicals) Regulation and the Drinking Water Directive. The REACH Regulation includes provisions for the restriction of certain PFAS substances, while the Drinking Water Directive sets limits for various contaminants, including PFAS. The EU is also working on a dedicated PFAS strategy, which aims to address the entire lifecycle of PFAS, from production to disposal.
- **Precautionary Principle:** The EU regulatory approach is grounded in the precautionary principle, which prioritises prevention and risk minimisation. This approach has led to proactive measures such as banning or restricting the use of specific PFAS compounds and investing in research to better understand their impacts.
- **Coordinated Efforts:** The EU coordinates its PFAS management efforts through various agencies, including the European Chemicals Agency (ECHA) and the European Food Safety Authority (EFSA). These agencies provide scientific assessments and recommendations that

inform regulatory decisions and policy development.

3. Australia: National and State-Based Strategies

Australia's approach to PFAS regulation involves a combination of national guidelines and state-based actions.

- **National PFAS Position Statement:** The Australian Government has issued a National PFAS Position Statement, outlining a framework for managing PFAS contamination. This framework includes principles for monitoring, risk assessment, and remediation. The Position Statement emphasises a precautionary approach and encourages collaboration between federal, state, and territory governments.

- **State and Territory Regulations:** Each state and territory in Australia has developed its own PFAS management strategies, reflecting regional priorities and conditions. For example, Victoria has implemented a PFAS Management Framework that includes guidelines for assessing and managing PFAS contamination. Queensland has established a PFAS Action Plan focusing on monitoring, remediation, and community engagement.

- **Research and Innovation:** Australia invests in research to improve PFAS detection, risk assessment, and remediation technologies. This research supports the development of effective strategies for managing PFAS contamination and informs regulatory decisions.

4. Comparative Insights: Key Lessons and Best Practices
Integrated Approach:

- **Holistic Management:** A comprehensive regulatory approach that integrates monitoring, prevention, and remediation is crucial. The EU's emphasis on the entire lifecycle of PFAS and the US's multi-layered strategy highlight the benefits of a holistic approach. The UK's regulatory framework could benefit from incorporating similar elements to address PFAS contamination more effectively.

Precautionary Principle:

- **Proactive Measures:** Adopting a precautionary approach, as seen in the EU, can help prevent PFAS contamination and minimise risks. Implementing stricter regulations and restrictions on PFAS use, combined with early intervention and research, can enhance the effectiveness of PFAS management in the UK.

Coordination and Collaboration:

- **Multi-Stakeholder Engagement:** Successful PFAS management requires collaboration among various stakeholders, including government agencies, industry, and communities. The coordinated efforts observed in the EU and Australia demonstrate the importance of involving multiple parties in addressing PFAS issues. The UK can strengthen its approach by enhancing coordination between regulatory bodies and engaging with affected communities.

Research and Innovation:

- **Continuous Improvement:** Investing in research and innovation is essential for developing new technologies and strategies for PFAS management. The US and Australia both highlight the importance of ongoing research to improve detection methods, remediation technologies, and risk assessment. The UK should continue to support research initiatives and adopt innovative solutions to address PFAS contamination.

5. Conclusion

Comparing the UK's PFAS regulatory framework with international practices reveals both strengths and areas for improvement. The diverse approaches adopted by the US, EU, and Australia provide valuable insights into effective PFAS management. By integrating lessons from these international experiences, the UK can enhance its regulatory framework, address PFAS contamination more comprehensively, and protect public health and the environment. Adopting best practices, emphasising precaution, and fostering collaboration will be key to advancing the UK's efforts in managing PFAS pollution and mitigating its impacts.

Future Directions and Recommendations for Enhancing PFAS Regulation in the UK

As awareness of the persistent and widespread nature of PFAS (Per- and Polyfluoroalkyl Substances) contamination continues to grow, it is imperative that the UK's regulatory framework evolves to address the challenges effectively. This section outlines potential future directions and recommendations for enhancing PFAS regulation in the UK, focusing on strengthening policies, improving enforcement, and fostering collaboration.

1. Strengthening Regulatory Policies

A. Comprehensive PFAS Strategy:

- **Integrated Framework:** The development of a comprehensive PFAS strategy should involve an integrated approach that encompasses prevention, monitoring, and remediation. This strategy should outline clear objectives, action plans, and timelines for addressing PFAS contamination. The UK could benefit from a national strategy similar to the EU's PFAS strategy, which addresses the entire lifecycle of PFAS from production to disposal.
- **Strict Regulations and Limits:** Revising and tightening

regulations to establish stricter limits for PFAS in water, soil, and air is crucial. The UK should consider aligning its standards with those of countries that have set stringent limits, such as the US and certain EU member states. Regular updates to these standards, based on emerging scientific evidence, will ensure they remain protective of human health and the environment.

B. Expansion of Regulated PFAS Compounds:

- **Broader Coverage:** Expanding the list of regulated PFAS compounds to include a wider range of substances is necessary. Current regulations often focus on a limited number of PFAS, leaving gaps in the management of other potentially harmful compounds. By broadening the scope, the UK can better address the diverse range of PFAS present in the environment.
- **Periodic Review:** Establishing a framework for the periodic review and inclusion of new PFAS compounds as they are identified will enhance regulatory effectiveness. This proactive approach ensures that emerging PFAS threats are addressed in a timely manner.

2. Enhancing Enforcement and Compliance
A. Improving Monitoring and Detection:

- **Advanced Analytical Techniques:** Investing in advanced analytical techniques and technologies for detecting low concentrations of PFAS is essential. Enhanced monitoring capabilities will improve the accuracy of contamination assessments and enable more effective enforcement of regulatory limits.

- **Expanded Monitoring Network:** Expanding the network of monitoring sites to cover a broader range of water sources, soils, and sediments will provide a more comprehensive understanding of PFAS contamination. Increased sampling frequency and spatial coverage will enhance the detection of contamination hotspots and support targeted remediation efforts.

B. Strengthening Enforcement Mechanisms:

- **Increased Resources:** Allocating additional resources for enforcement activities, including funding, personnel, and training, will strengthen the ability of regulatory agencies to ensure compliance. This includes increasing the capacity for inspections, audits, and investigations into non-compliance.
- **Clear Penalties and Incentives:** Implementing clear and enforceable penalties for violations of PFAS regulations, along with incentives for compliance, will encourage adherence to regulatory requirements. A well-defined framework for addressing non-compliance and rewarding proactive measures can enhance overall enforcement effectiveness.

3. Promoting Research and Innovation
A. Supporting Research Initiatives:

- **Funding and Collaboration:** Increasing funding for research on PFAS detection, risk assessment, and remediation technologies is crucial. Collaborating with academic institutions, research organisations, and industry experts can drive innovation and lead to the development of more effective solutions for managing PFAS contamination.

- **Knowledge Sharing:** Promoting the sharing of research findings and best practices among countries and organisations will facilitate the development of cutting-edge technologies and strategies. Engaging in international research collaborations can provide valuable insights and accelerate progress in PFAS management.

B. Developing New Technologies:

- **Innovative Remediation Technologies:** Investing in the development and implementation of innovative remediation technologies, such as advanced oxidation processes, bioremediation, and novel filtration methods, will enhance the ability to address PFAS contamination effectively. Pilot projects and demonstration studies can help validate the efficacy of new technologies before widespread adoption.
- **Enhanced Detection Methods:** Improving detection methods for PFAS, including portable and real-time analytical tools, will enable more efficient monitoring and rapid response to contamination incidents. The development of user-friendly and cost-effective detection technologies will support both regulatory agencies and industry stakeholders.

4. Fostering Collaboration and Public Engagement
A. Engaging Stakeholders:

- **Multi-Stakeholder Approach:** Adopting a multi-stakeholder approach that involves government agencies, industry, environmental organisations, and communities will foster collaboration and improve PFAS management. Engaging stakeholders in the development of policies and strategies ensures that diverse perspectives and expertise are considered.

- **Community Involvement:** Involving affected communities in decision-making processes and providing transparent information about PFAS contamination and remediation efforts will build trust and support. Public engagement initiatives, such as community meetings and information campaigns, can help address concerns and promote informed decision-making.

B. International Cooperation:

- **Global Standards and Agreements:** Participating in international efforts to develop global standards and agreements for PFAS management will enhance the UK's regulatory framework. Collaboration with international organisations and other countries can lead to the harmonisation of regulations and the sharing of best practices.
- **Joint Research and Initiatives:** Engaging in joint research initiatives and collaborative projects with other countries can advance the understanding of PFAS impacts and improve management strategies. International cooperation can accelerate progress and provide valuable insights into effective PFAS regulation.

5. Conclusion

The future direction of PFAS regulation in the UK requires a multifaceted approach that addresses policy, enforcement, research, and collaboration. Strengthening regulations, enhancing enforcement, promoting research and innovation, and fostering stakeholder engagement are key to improving the management of PFAS contamination. By adopting these recommendations and learning from international practices, the UK can develop a more effective regulatory framework that safeguards public health and the environment from the impacts of PFAS pollution. As the understanding of PFAS continues to evolve, ongoing adaptation and improvement of regulatory measures will be essential in addressing this complex and persistent environmental challenge.

The Role of Industry – Responsibility and Accountability

The Role of Industries in PFAS Pollution

Industries have played a significant role in the proliferation of PFAS (Per- and Polyfluoroalkyl Substances) pollution due to their widespread use of these chemicals in various applications. Understanding how different sectors contribute to PFAS contamination and the steps they have taken—or failed to take—in addressing these issues is crucial for developing effective solutions and ensuring corporate accountability.

1. Industrial Use of PFAS

A. Historical Applications:

- **Firefighting Foams:** One of the most well-documented sources of PFAS pollution comes from the use of firefighting foams, particularly in airports, military bases, and industrial sites. These foams contain PFAS compounds designed to suppress flammable liquid fires. Although their effectiveness in emergency situations is high, their environmental impact is severe. The persistent nature of PFAS means that they can leach into soil and groundwater, leading to long-term contamination.

- **Manufacturing and Industrial Processes:** PFAS have been used extensively in manufacturing processes, including the production of non-stick cookware, stain-resistant fabrics, and water-repellent materials. The chemicals provide desirable properties such as resistance to heat, stains, and water, which have made them popular in a wide range of consumer products. However, the industrial handling and disposal of these substances have contributed to widespread environmental contamination.

- **Chemical Production:** Companies involved in the production of PFAS have historically discharged waste containing these substances into landfills, water bodies, and the air. The legacy of such practices has left a lasting impact on many industrial sites and surrounding communities. These discharges, combined with inadequate waste management practices, have resulted in substantial PFAS contamination.

B. Regulatory Gaps and Industry Practices:

- **Lack of Early Regulation:** For many years, the use and disposal of PFAS were poorly regulated, allowing industries to use these chemicals with minimal oversight. The lack of early regulatory controls meant that companies did not face stringent requirements for managing waste or mitigating environmental impacts. This regulatory gap contributed to the widespread dispersion of PFAS across different environmental media.

- **Inadequate Waste Management:** Historically, many industries did not implement adequate waste management practices for PFAS-containing materials. In some cases, waste was disposed of in ways that led to leaching into soil and groundwater. The use of PFAS in various industrial processes often lacked sufficient controls to prevent environmental release, exacerbating the problem.

2. Case Studies of PFAS-Related Contamination
A. Case Study: 3M Company

- **Background:** 3M, a major manufacturer of PFAS-containing products, has faced significant scrutiny over its role in PFAS pollution. The company produced a range of PFAS chemicals, including those used in firefighting foams and consumer products like Scotchgard. As evidence of PFAS-related health and environmental impacts grew, 3M came under pressure to address its role in the contamination.

- **Response and Actions:** In response to mounting evidence and regulatory pressure, 3M began phasing out the production of certain PFAS compounds and committed to reducing its environmental impact. The company has also been involved in litigation and settlements related to PFAS contamination. 3M's actions highlight the challenges and complexities of addressing legacy contamination and transitioning to more sustainable practices.

B. Case Study: DuPont and Chemours

- **Background:** DuPont, another major player in the PFAS industry, has been linked to significant PFAS contamination through its production of chemicals like PFOA (Perfluorooctanoic Acid). The contamination associated with DuPont's operations has affected communities around its manufacturing sites and led to numerous legal and regulatory actions.
- **Response and Actions:** DuPont has faced extensive legal battles and has been required to pay substantial settlements related to PFAS contamination. In 2015, DuPont spun off its chemical operations into a new company, Chemours, which has also faced legal challenges related to PFAS. The ongoing legal and financial consequences for both

companies underscore the long-term implications of PFAS pollution.

3. Corporate Responsibility and Sustainable Practices
A. Shifting Towards Sustainability:

- **Adopting Green Chemistry:** Many companies are increasingly embracing green chemistry principles to reduce the use of hazardous substances and minimise environmental impacts. This shift involves developing alternative materials and processes that do not rely on PFAS or other harmful chemicals. Green chemistry practices aim to prevent pollution at the source and promote the use of safer chemicals in industrial processes.

- **Implementing Circular Economy Principles:** The concept of a circular economy focuses on designing products and processes that minimise waste and maximise the reuse and recycling of materials. By adopting circular economy principles, companies can reduce their reliance on PFAS and other persistent pollutants, thereby mitigating their environmental footprint.

B. Enhancing Transparency and Accountability:

- **Disclosure and Reporting:** Companies are increasingly expected to provide transparency about their environmental impacts and chemical usage. Enhanced disclosure requirements and reporting standards can help stakeholders understand the extent of a company's PFAS-related activities and their environmental impact. Transparent reporting fosters accountability and encourages companies to adopt more sustainable practices.

- **Community Engagement:** Engaging with affected communities and addressing their concerns is a crucial aspect of corporate responsibility. Companies involved in PFAS pollution should work collaboratively with local communities to provide information, support remediation efforts, and address health and environmental concerns. Building trust and maintaining open communication channels are essential for effective community relations.

C. Regulatory Compliance and Beyond:

- **Proactive Measures:** Companies should not only comply with existing regulations but also take proactive measures to prevent future contamination. This includes investing in research to identify safer alternatives to PFAS, implementing best practices for waste management, and supporting initiatives aimed at reducing environmental contamination.
- **Voluntary Initiatives:** Some companies are voluntarily adopting more stringent environmental standards and practices beyond regulatory requirements. Participating in industry-wide initiatives and certifications, such as those related to environmental sustainability and chemical safety, can demonstrate a commitment to responsible practices and enhance a company's reputation.

4. Challenges and Opportunities
A. Addressing Legacy Contamination:

- **Remediation and Cleanup:** Addressing legacy PFAS contamination presents significant challenges, including the need for effective remediation technologies and strategies. Companies involved in historical contamination must invest in cleanup efforts and collaborate with regulatory agencies and communities to mitigate the impacts of past practices.

- **Legal and Financial Implications:** The financial and legal consequences of PFAS contamination can be substantial. Companies may face lawsuits, regulatory fines, and reputational damage. Managing these implications requires a proactive approach to addressing contamination and engaging with stakeholders.

B. Promoting Innovation:

- **Research and Development:** Investing in research and development is essential for finding innovative solutions to PFAS contamination. Companies that lead in developing new technologies and safer alternatives can contribute to advancing environmental protection and setting industry standards.
- **Collaborative Efforts:** Collaboration between industry, government, and research institutions can drive progress in managing PFAS pollution. Joint initiatives and partnerships can facilitate the sharing of knowledge and resources, leading to more effective solutions and regulatory approaches.

5. Conclusion

Industries have played a pivotal role in the proliferation of PFAS pollution, with significant impacts on the environment and public health. Examining the historical use of PFAS, case studies of major companies, and the evolving landscape of corporate responsibility highlights the need for enhanced accountability and sustainable practices. By addressing legacy contamination, adopting green chemistry principles, and fostering transparency, companies can contribute to mitigating the impacts of PFAS pollution and promoting a more sustainable future. The ongoing challenge is to ensure that industries take meaningful actions to prevent future contamination and support efforts to protect both human health and the environment.

Case Studies of Companies Involved in PFAS Production and Their Responses to Contamination Issues

Examining the responses of companies involved in PFAS production provides insight into how the industrial sector has addressed PFAS-related contamination issues. By reviewing specific case studies, we can assess the effectiveness of corporate actions, the challenges they faced, and the lessons learned. This analysis highlights the complexities of managing PFAS contamination and the evolving nature of corporate responsibility.

1. Case Study: 3M Company

A. Background:

3M Company, a major manufacturer of PFAS-containing products, played a significant role in the widespread use of these chemicals. The company produced a range of PFAS compounds, including those used in firefighting foams, stain-resistant treatments, and non-stick coatings. 3M's involvement in PFAS production has had far-reaching environmental and health implications.

B. Response and Actions:

- **Phase-Out of PFAS Production:** In response to growing evidence of PFAS-related health risks and environmental contamination, 3M announced in 2000 that it would phase out the production of perfluorooctane sulfonate (PFOS), a specific type of PFAS. The decision followed increasing scrutiny from regulatory agencies and mounting public concern about the chemical's impact.

- **Settlement and Remediation:** 3M has faced numerous lawsuits related to PFAS contamination. In 2018, the company reached a settlement with the State of Minnesota, agreeing to pay $850 million to address contamination from its PFAS products. The settlement included funds for environmental remediation and health studies. Additionally, 3M has been involved in various cleanup efforts at contaminated sites and has committed to improving its environmental stewardship.

- **Corporate Commitments:** 3M has made several commitments to improve its environmental performance, including pledges to reduce its carbon footprint and advance sustainable practices. The company has also focused on developing safer alternatives to PFAS in its product lines.

2. Case Study: DuPont and Chemours
A. Background:

DuPont, another major player in the PFAS industry, was a significant producer of perfluorooctanoic acid (PFOA) and other PFAS chemicals. The company's operations resulted in substantial PFAS contamination at various sites, leading to extensive legal and regulatory challenges. In 2015, DuPont spun off its chemical operations into a new company, Chemours, which continued to face scrutiny over PFAS issues.

B. Response and Actions:

- **Legal Settlements:** DuPont has been involved in numerous legal battles related to PFAS contamination. In 2017, DuPont and Chemours agreed to a settlement of $671 million to resolve claims related to PFOA contamination. This settlement covered remediation efforts and compensation for affected communities. DuPont has also faced individual lawsuits and has been required to pay damages for health-related claims.

- **Transition to Alternative Chemicals:** Both DuPont and Chemours have committed to reducing their use of harmful PFAS chemicals and transitioning to alternative substances. Chemours, in particular, has focused on developing products with reduced environmental and health impacts. The companies have also worked to improve their waste management practices and reduce emissions from production processes.

- **Transparency and Reporting:** DuPont has taken steps to enhance transparency regarding its environmental performance and chemical usage. The company has increased its reporting on environmental impacts and has worked to address concerns from affected communities. However, critics argue that these efforts are not sufficient to fully address the legacy of PFAS contamination.

3. Case Study: Aqueous Film-Forming Foam (AFFF) Manufacturers

A. Background:

Aqueous Film-Forming Foam (AFFF) is a firefighting foam that contains high levels of PFAS compounds. Several manufacturers have been involved in the production and distribution of AFFF, including companies like Tyco Fire Products and Chemguard. The widespread use of AFFF in firefighting has resulted in significant PFAS contamination at many military bases, airports, and industrial sites.

B. Response and Actions:

- **Litigation and Settlements:** Manufacturers of AFFF have faced numerous lawsuits related to PFAS contamination. In recent years, several companies have reached settlements to address claims from affected parties. These settlements often include funds for environmental cleanup, health studies, and compensation for affected communities.
- **Product Reformulation:** In response to legal and regulatory pressures, some AFFF manufacturers have begun reformulating their products to reduce or eliminate PFAS content. The shift towards PFAS-free alternatives aims to minimise the environmental impact of firefighting foams and address concerns about health risks associated with traditional AFFF products.
- **Industry Collaboration:** The firefighting foam industry has seen increased collaboration among manufacturers, regulators, and environmental groups to develop and implement safer alternatives. Efforts include the development of new foam formulations and improved waste management practices to reduce the environmental footprint of firefighting activities.

4. Corporate Responsibility and Challenges
A. Addressing Legacy Contamination:

- **Remediation Efforts:** Companies involved in PFAS production face significant challenges in addressing legacy contamination. Remediation efforts often require substantial investment and time to effectively clean up contaminated sites. Companies must navigate complex regulatory requirements and collaborate with stakeholders to manage and mitigate the impacts of past practices.

- **Financial and Reputational Costs:** The financial and reputational costs of PFAS contamination can be considerable. Companies may face legal liabilities, regulatory fines, and damage to their public image. Addressing these costs requires a comprehensive approach that includes remediation, transparency, and proactive measures to prevent future contamination.

B. Moving Towards Sustainable Practices:

- **Adopting Green Chemistry:** Companies are increasingly adopting green chemistry principles to reduce their reliance on hazardous substances like PFAS. Green chemistry focuses on designing products and processes that minimise environmental and health impacts. By investing in research and development, companies can identify and implement safer alternatives to PFAS.
- **Enhancing Transparency:** Transparency in environmental reporting and chemical usage is essential for building trust with stakeholders and demonstrating corporate responsibility. Companies that proactively disclose information about their environmental performance and chemical management practices can enhance their reputation and address concerns about PFAS contamination.

5. Conclusion

The case studies of companies involved in PFAS production illustrate the complex and evolving nature of corporate responsibility and accountability. While some companies have taken significant steps to address PFAS-related contamination, challenges remain in managing legacy issues and transitioning to more sustainable practices. By examining these case studies, we gain valuable insights into the effectiveness of corporate actions and the need for continued efforts to mitigate the impact of PFAS pollution. Moving forward, a focus on remediation, transparency, and sustainable practices will be essential for addressing the challenges of PFAS contamination and promoting a healthier environment.

Corporate Responsibility and the Push for More Sustainable Practices

The escalating awareness of PFAS (Per- and Polyfluoroalkyl Substances) contamination has heightened scrutiny of the role played by industries in addressing these issues. Corporate responsibility is increasingly under the spotlight, with expectations for companies to adopt more sustainable practices, ensure transparency, and mitigate their environmental impact. This section delves into how industries are responding to the challenges posed by PFAS pollution and the evolving standards for corporate responsibility.

1. Embracing Sustainable Practices

A. Green Chemistry and Innovation:

- **Principles of Green Chemistry:** Green chemistry focuses on designing chemical products and processes that reduce or eliminate the use and generation of hazardous substances. For companies involved in PFAS production, adopting green chemistry principles is essential. This involves developing alternatives that are safer for human health and the environment, reducing the reliance on harmful PFAS chemicals.

- **Case Examples of Innovation:** Companies are increasingly investing in research and development to find safer alternatives to PFAS. For example, some manufacturers are exploring the use of fluorine-free foams in place of PFAS-containing firefighting foams. These alternatives aim to provide the necessary performance without the environmental persistence and health risks associated with traditional PFAS products.

- **Product Reformulation:** Reformulating products to eliminate or reduce PFAS content is a key strategy for companies aiming to enhance sustainability. This process involves rigorous testing and validation to ensure that the new formulations meet performance standards while mitigating environmental and health impacts.

B. Circular Economy Initiatives:

- **Design for Sustainability:** The concept of a circular economy focuses on designing products and processes that minimise waste and promote the recycling and reuse of materials. By incorporating circular economy principles, companies can reduce their environmental footprint and manage PFAS-related waste more effectively.
- **Recycling and Disposal:** Effective recycling and disposal practices are crucial for managing PFAS-containing products. Companies are exploring innovative methods for recycling materials and treating waste to prevent PFAS contamination from spreading. This includes developing advanced technologies for treating contaminated soil and water and ensuring proper disposal of PFAS-containing products.

2. Enhancing Transparency and Accountability
A. Environmental Reporting and Disclosure:

- **Increased Reporting Requirements:** Companies are

facing growing pressure to provide transparent reports on their environmental performance, including the use and management of PFAS chemicals. Enhanced reporting requirements help stakeholders understand the extent of a company's environmental impact and its efforts to address contamination.

- **Voluntary Disclosures:** Many companies are going beyond regulatory requirements by voluntarily disclosing information about their environmental practices and chemical usage. This transparency can build trust with consumers, regulators, and communities, demonstrating a commitment to responsible environmental stewardship.

- **Sustainability Reports:** Comprehensive sustainability reports are becoming standard practice for companies committed to addressing environmental issues. These reports often include details on PFAS management, remediation efforts, and progress towards sustainability goals. Such transparency can enhance a company's reputation and accountability.

B. Engaging with Stakeholders:

- **Community Engagement:** Engaging with communities affected by PFAS contamination is a critical aspect of corporate responsibility. Companies are increasingly involved in dialogue with local residents, addressing their concerns, and supporting community-based initiatives to mitigate the impacts of contamination.

- **Collaborative Efforts:** Collaboration with environmental organisations, regulatory agencies, and other stakeholders is essential for effective PFAS management. Companies are working together with these groups to share knowledge, develop best practices, and implement strategies for reducing PFAS contamination.

- **Consumer Education:** Educating consumers about the risks associated with PFAS and the steps companies are taking to address these issues can foster a more informed and engaged public. Transparency in communication helps build trust and demonstrates a company's commitment to addressing environmental and health concerns.

3. Regulatory Compliance and Beyond
A. Meeting and Exceeding Regulations:

- **Compliance with Regulations:** Adhering to existing regulations is a fundamental aspect of corporate responsibility. Companies involved in PFAS production must comply with legal requirements related to chemical management, waste disposal, and environmental protection. Ensuring compliance helps prevent further contamination and demonstrates a commitment to regulatory standards.
- **Exceeding Regulatory Requirements:** In addition to meeting regulatory requirements, companies are increasingly adopting practices that exceed minimum standards. This proactive approach includes investing in advanced technologies, setting ambitious environmental goals, and participating in industry-wide initiatives aimed at reducing PFAS-related impacts.

B. Proactive Measures and Industry Leadership:

- **Setting Industry Standards:** Leading companies are

setting new standards for environmental performance and sustainability within their sectors. By adopting best practices and demonstrating leadership in PFAS management, these companies can influence industry trends and encourage others to follow suit.

- **Investing in Research and Development:** Proactively investing in research and development is crucial for advancing PFAS management practices. Companies that lead in developing innovative solutions and safer alternatives contribute to the broader effort to address PFAS contamination and enhance environmental sustainability.

- **Participating in Industry Initiatives:** Engaging in industry initiatives and collaborative projects helps drive progress in managing PFAS pollution. By working with other companies, regulatory agencies, and environmental groups, companies can contribute to the development of effective strategies and solutions for PFAS management.

4. Challenges and Opportunities
A. Managing Legacy Contamination:

- **Remediation and Cleanup Efforts:** Addressing legacy PFAS contamination presents significant challenges, including the need for effective remediation technologies and strategies. Companies involved in historical contamination must invest in cleanup efforts, collaborate with regulators, and work with affected communities to manage and mitigate the impacts of past practices.

- **Financial Implications:** The financial implications of PFAS contamination can be substantial, including costs related to remediation, legal liabilities, and reputational damage. Managing these costs requires a comprehensive approach that includes effective remediation, transparent communication, and proactive measures to prevent future contamination.

B. Driving Innovation and Collaboration:

- **Advancing Technologies:** Investing in the development of new technologies and solutions for PFAS management is crucial for addressing the challenges posed by these chemicals. Companies that lead in innovation can drive progress and contribute to more effective and sustainable approaches to PFAS pollution.
- **Collaborative Approaches:** Collaboration between industry, government, and research institutions is essential for advancing PFAS management practices. Joint initiatives and partnerships can facilitate the sharing of knowledge and resources, leading to more effective solutions and regulatory approaches.

5. Conclusion

Corporate responsibility and accountability in the context of PFAS pollution are increasingly important as the environmental and health impacts of these chemicals become more apparent. By embracing sustainable practices, enhancing transparency, and proactively addressing legacy contamination, companies can contribute to mitigating the impacts of PFAS pollution and promoting a healthier environment. The challenges of managing PFAS contamination are significant, but through innovation, collaboration, and commitment to responsible practices, industries can play a pivotal role in addressing this complex issue. Moving forward, continued efforts to improve environmental performance and adopt sustainable practices will be essential for ensuring a safer and more sustainable future.

The Push for Policy Changes and the Role of Advocacy

As the awareness of PFAS (Per- and Polyfluoroalkyl Substances) contamination grows, there is increasing pressure on industries to adopt more sustainable practices and address the impacts of their operations. This part explores the role of advocacy and policy changes in driving industry responsibility, the influence of public opinion, and the steps being taken to reform practices and ensure accountability.

1. The Influence of Advocacy and Public Pressure

A. Role of Environmental Advocacy Groups:

- **Raising Awareness:** Environmental advocacy groups play a crucial role in highlighting the dangers of PFAS contamination and pushing for stricter regulations. Organisations such as Greenpeace, the Environmental Working Group (EWG), and local environmental groups have been instrumental in bringing PFAS issues to the forefront of public consciousness. Their campaigns have educated the public about the risks associated with PFAS and have pressured companies to take action.
- **Legal Actions and Campaigns:** Advocacy groups often engage in legal actions and public campaigns to hold companies accountable for PFAS pollution. Lawsuits and petitions can compel companies to disclose information about their PFAS practices, undertake remediation efforts, and compensate affected communities. These actions create a sense of urgency and drive industries to address contamination more effectively.
- **Influencing Policy Changes:** Environmental groups

advocate for stronger regulations and policies related to PFAS. Their efforts include lobbying for legislative changes, supporting research into safer alternatives, and promoting stricter enforcement of environmental standards. Advocacy can lead to more comprehensive regulations and increased corporate accountability.

B. Public Opinion and Consumer Activism:

- **Consumer Demands:** Public awareness of PFAS contamination has led to increased consumer demands for safer products and more transparent corporate practices. Consumers are increasingly seeking products that do not contain PFAS or other harmful chemicals, influencing companies to reformulate their products and adopt more sustainable practices.
- **Boycotts and Protests:** Consumer activism can manifest in boycotts and protests aimed at companies involved in PFAS contamination. Such actions can have a significant impact on a company's reputation and financial performance, prompting them to address environmental and health concerns more proactively.
- **Support for Policy Reform:** Public pressure also supports efforts to reform policies and regulations related to PFAS. Campaigns advocating for stronger government action can lead to more stringent regulations and increased investment in research and remediation efforts. The collective voice of concerned citizens can drive meaningful change in industry practices and regulatory frameworks.

2. Policy Changes and Regulatory Reforms
A. Current Regulatory Landscape:

- **Existing Regulations:** In the UK, regulatory frameworks for managing PFAS are evolving. The Environment Agency and other regulatory bodies have established guidelines for the management of PFAS-contaminated sites and the use of PFAS in products. However, these regulations are often criticised for being insufficiently comprehensive or for lagging behind scientific understanding of PFAS risks.

- **European Union Initiatives:** The EU has implemented several initiatives to address PFAS contamination, including restrictions on the use of certain PFAS compounds and requirements for monitoring and reporting. The European Chemicals Agency (ECHA) has been actively involved in assessing the risks of PFAS and proposing regulatory measures to manage their impact.

- **Legislative Developments:** Recent legislative developments in the UK and the EU aim to strengthen regulations around PFAS. These include proposals to restrict the use of certain PFAS compounds, enhance monitoring and reporting requirements, and support research into safer alternatives. Such changes reflect a growing recognition of the need for more robust regulatory measures.

B. Policy Reform and Industry Response:

- **Strengthening Regulations:** The push for stronger regulations is driven by the need to address the persistent and widespread nature of PFAS contamination. Proposed reforms include lowering acceptable limits for PFAS in water and soil, expanding the scope of regulatory oversight, and increasing penalties for non-compliance. These measures aim to reduce the release of PFAS into the environment and protect public health.

- **Industry Adaptation:** In response to regulatory changes, industries are adapting their practices to comply with new standards. This includes reformulating products, improving waste management practices, and investing in technologies for PFAS removal and remediation. Companies that proactively address regulatory requirements may gain a competitive advantage and enhance their reputation for environmental stewardship.

- **Collaborative Efforts:** Effective policy reform often requires collaboration between regulators, industry stakeholders, and advocacy groups. Joint initiatives can facilitate the development of practical and enforceable regulations, provide support for innovation, and ensure that regulatory measures are implemented effectively. Collaborative approaches can help address the complex challenges associated with PFAS contamination.

3. The Future of PFAS Management
A. Advancing Research and Innovation:

- **Development of Alternatives:** Research into safer alternatives to PFAS is a critical component of future PFAS management. Investment in innovative technologies and materials that do not rely on PFAS can help reduce the environmental impact of industrial processes and consumer products. Continued research is necessary to identify effective and sustainable alternatives.

- **Improving Remediation Technologies:** Advancements in

remediation technologies are essential for addressing existing PFAS contamination. Research into more efficient and cost-effective methods for removing PFAS from contaminated soil and water can support cleanup efforts and mitigate the impacts of legacy pollution.

- **Enhancing Monitoring and Detection:** Improved monitoring and detection methods are crucial for identifying and managing PFAS contamination. Advances in analytical techniques and monitoring technologies can enhance the ability to detect PFAS at lower concentrations and in a wider range of environmental media.

B. Building a Sustainable Future:

- **Integrating Sustainability:** Integrating sustainability principles into industrial practices and product design can help prevent future PFAS contamination. Companies that prioritise environmental and health considerations in their operations contribute to a more sustainable future and reduce the risk of harmful substances entering the environment.

- **Encouraging Responsible Practices:** Encouraging responsible practices across industries can drive progress in managing PFAS contamination. This includes promoting transparency, supporting regulatory reforms, and fostering collaboration between stakeholders. A collective effort to address PFAS challenges can lead to more effective solutions and a healthier environment.

- **Public and Industry Engagement:** Engaging both the public and industry stakeholders in efforts to manage PFAS contamination is essential for achieving meaningful change. Public awareness and advocacy can drive regulatory reforms and corporate responsibility, while industry participation in research and innovation can contribute to sustainable solutions.

4. Conclusion

The push for policy changes and the role of advocacy are critical in shaping the future of PFAS management and industry responsibility. As awareness of PFAS contamination grows, there is increasing pressure on companies to adopt more sustainable practices, comply with evolving regulations, and address the impacts of their operations. Through advocacy, policy reform, and collaborative efforts, significant progress can be made in managing PFAS contamination and protecting public health and the environment. The future of PFAS management will depend on continued innovation, responsible practices, and a collective commitment to addressing one of the most pressing environmental challenges of our time.

Advances in PFAS Detection and Remediation Technologies

Recent Technological Advancements in Detecting PFAS in Water Sources

PFAS (Per- and Polyfluoroalkyl Substances) contamination poses a significant challenge due to the chemicals' persistence and low detection limits. Recent advancements in detection technologies have improved our ability to identify and measure PFAS in water sources with greater accuracy and sensitivity. This section reviews these technological advancements, their implications for environmental monitoring, and the ongoing efforts to enhance detection methods.

1. Enhanced Analytical Techniques

A. High-Resolution Mass Spectrometry (HRMS):

- **Principles and Advantages:** High-resolution mass spectrometry (HRMS) has revolutionised PFAS detection by providing unparalleled sensitivity and specificity. HRMS techniques, such as quadrupole time-of-flight (QTOF) and Orbitrap mass spectrometers, allow for the precise identification of PFAS compounds at very low concentrations. These instruments can differentiate between various PFAS isomers and congeners, improving the accuracy of analyses.

- **Applications in PFAS Detection:** HRMS is particularly effective in identifying trace levels of PFAS in complex environmental matrices, such as water and soil samples. The technique supports the detection of a wide range of PFAS compounds, including those that are challenging to analyse using traditional methods. HRMS enables detailed profiling of PFAS contamination, aiding in source identification and risk assessment.

B. Liquid Chromatography-Mass Spectrometry (LC-MS):

- **Advanced LC-MS Techniques:** Liquid chromatography-mass spectrometry (LC-MS) continues to be a cornerstone in PFAS detection. Recent advancements in LC-MS technology, including the development of high-resolution and high-sensitivity systems, have enhanced the ability to detect and quantify PFAS compounds. Improvements in chromatographic separation and ionisation techniques contribute to more accurate and reliable results.

- **Targeted and Non-Targeted Approaches:** LC-MS can be used in both targeted and non-targeted approaches for PFAS detection. Targeted LC-MS methods focus on specific PFAS compounds of interest, providing high sensitivity and selectivity. Non-targeted approaches, such as suspect screening and untargeted analysis, allow for the identification of unknown or unexpected PFAS compounds, broadening the scope of detection.

C. Screening Technologies:

- **Immunoassay Methods:** Immunoassay techniques, such as enzyme-linked immunosorbent assays (ELISA), offer a rapid and cost-effective approach for screening PFAS in water samples. These methods use antibodies to detect specific PFAS compounds or classes of PFAS, providing a preliminary assessment of contamination levels.

- **Portable Analytical Devices:** Advances in portable analytical devices, including field deployable LC-MS and

ELISA systems, enable real-time monitoring of PFAS in remote or contaminated areas. These devices facilitate on-site analysis, reducing the time and cost associated with laboratory-based testing.

2. Improved Detection Sensitivity
A. Low Detection Limits:

- **Trace Analysis Capabilities:** Recent advancements have lowered the detection limits for PFAS, allowing for the measurement of compounds at parts per trillion (ppt) levels. This increased sensitivity is crucial for detecting low concentrations of PFAS in drinking water, groundwater, and surface water, where regulatory limits are often stringent.
- **Method Validation and Standardisation:** The development of validated analytical methods and standardisation practices enhances the reliability of PFAS detection. International standardisation organisations, such as the American Society for Testing and Materials (ASTM) and the International Organisation for Standardisation (ISO), provide guidelines and protocols for PFAS analysis, ensuring consistency and accuracy across different laboratories.

B. Matrix Effects and Sample Preparation:

- **Overcoming Matrix Interference:** Advances in sample preparation techniques, such as solid-phase extraction (SPE) and liquid-liquid extraction (LLE), help to overcome matrix interference in environmental samples. These techniques improve the purity of extracted PFAS compounds and enhance the accuracy of detection.

- **Enhanced Sample Preservation:** Innovations in sample preservation methods, such as the use of stabilising agents and improved storage conditions, reduce the potential for PFAS degradation during sample handling and transportation. Proper preservation is essential for maintaining the integrity of samples and ensuring accurate analytical results.

3. Emerging Technologies
A. Nanotechnology in PFAS Detection:

- **Nanomaterial-Based Sensors:** Nanotechnology is emerging as a powerful tool for PFAS detection. Nanomaterial-based sensors, such as gold nanoparticles and carbon nanotubes, offer high sensitivity and selectivity for PFAS compounds. These sensors can be integrated into portable devices for on-site monitoring and provide rapid and accurate results.

- **Magnetic Nanoparticles:** Magnetic nanoparticles can be used in combination with separation techniques to enhance PFAS detection. These nanoparticles selectively capture PFAS compounds, allowing for their concentration and subsequent analysis using various detection methods.

B. Biosensors and Biorecognition:

- **Biomolecular Detection:** Biosensors that utilise biomolecular recognition elements, such as antibodies or aptamers, offer a specific and sensitive approach to PFAS detection. These biosensors can be designed to detect individual PFAS compounds or classes of PFAS, providing a targeted analysis of contamination.

- **Microbial Biosensors:** Microbial biosensors, which use genetically engineered microorganisms to detect PFAS, represent an innovative approach to environmental monitoring. These biosensors can provide real-time information about PFAS contamination and are adaptable for various environmental conditions.

4. Integration and Data Management
A. Data Integration and Interpretation:

- **Data Fusion Techniques:** Integrating data from multiple detection methods enhances the accuracy and reliability of PFAS monitoring. Data fusion techniques combine results from HRMS, LC-MS, and other analytical methods to provide a comprehensive assessment of PFAS contamination.

- **Advanced Data Analytics:** The use of advanced data analytics, including machine learning and artificial intelligence, can improve the interpretation of complex PFAS data. These technologies help identify patterns, trends, and correlations in contamination data, supporting more informed decision-making.

B. Remote Sensing and Monitoring Networks:

- **Deployment of Sensor Networks:** The development of sensor networks for continuous monitoring of PFAS in water sources is an area of active research. These networks provide real-time data on PFAS levels and enable early detection of contamination events.

- **Remote Sensing Technologies:** Remote sensing technologies, such as satellite and aerial imaging, can be used to assess large-scale environmental contamination. These technologies offer valuable insights into PFAS distribution and movement in water bodies.

5. Conclusion

Recent advancements in PFAS detection technologies have significantly improved our ability to identify and measure these persistent contaminants in water sources. Enhanced analytical techniques, increased sensitivity, and emerging technologies contribute to more accurate and comprehensive monitoring of PFAS contamination. As detection methods continue to evolve, they will play a crucial role in addressing the challenges posed by PFAS pollution and supporting efforts to protect public health and the environment. Future developments in detection technologies, combined with effective remediation strategies, will be essential for managing PFAS contamination and ensuring a sustainable and safe water supply.

Analysis of Current Remediation Techniques and Their Effectiveness

Addressing PFAS (Per- and Polyfluoroalkyl Substances) contamination requires effective remediation techniques capable of managing these persistent chemicals. As PFAS are notoriously difficult to degrade and remove, various methods have been developed and are being evaluated for their efficacy. This section provides an in-depth analysis of current remediation techniques, their effectiveness, limitations, and the ongoing efforts to enhance these methods.

1. Conventional Remediation Techniques

A. Activated Carbon Filtration:

- **Mechanism and Applications:** Activated carbon filtration is one of the most widely used methods for treating PFAS-contaminated water. The process relies on the adsorption of PFAS molecules onto the surface of activated carbon. This technique is effective for removing a range of PFAS compounds, particularly those with shorter carbon chains.
- **Effectiveness and Limitations:** Activated carbon is effective in reducing PFAS concentrations to below regulatory limits in many cases. However, its effectiveness can diminish over time as the carbon becomes saturated with PFAS, necessitating regular replacement or regeneration. Additionally, activated carbon may not be as effective for longer-chain PFAS or for the complete removal of all PFAS compounds, as some may be only partially adsorbed.

B. Ion Exchange:

- **Process Overview:** Ion exchange involves the use of resins that exchange PFAS ions in contaminated water with less harmful ions. This method is particularly effective for certain PFAS types, including perfluorosulfonic acids (PFSAs) and perfluorocarboxylic acids (PFCAs).
- **Effectiveness and Constraints:** Ion exchange resins can be highly effective for reducing PFAS levels in water. However, the resins can become saturated and require periodic regeneration or replacement. The disposal of spent resins poses environmental and financial challenges, as they may become a concentrated source of PFAS that needs further treatment.

C. Reverse Osmosis:

- **Description and Mechanism:** Reverse osmosis (RO) is a filtration method that uses a semi-permeable membrane to remove contaminants from water. The process involves applying pressure to force water through the membrane, leaving PFAS and other contaminants behind.
- **Performance and Drawbacks:** Reverse osmosis is highly effective in removing a wide range of PFAS compounds, including those with longer carbon chains. It can achieve very low PFAS concentrations. However, RO systems are energy-intensive and generate a concentrate stream that contains high levels of PFAS, which requires additional treatment or disposal.

2. Advanced Remediation Technologies
A. Destruction Technologies:

- **High-Temperature Incineration:** High-temperature incineration involves burning PFAS-containing materials at very high temperatures to break down the chemicals. This method can effectively destroy PFAS, converting them into less harmful substances such as carbon dioxide and hydrogen fluoride.
- **Effectiveness and Concerns:** Incineration is effective at reducing PFAS concentrations to non-detectable levels. However, it requires specialised facilities capable of reaching the necessary temperatures and managing the by-products, including potentially hazardous gases. The high cost and environmental impact of incineration make it less practical for widespread use.
- **Plasma Arc Technology:** Plasma arc technology uses a high-energy plasma to break down PFAS compounds into

their basic elements. This method is still under development but shows promise in effectively destroying PFAS with high efficiency. Challenges include the high cost of plasma equipment and the need for further research to optimise the technology for broader application.

B. Chemical Oxidation and Reduction:

- **Oxidative Processes:** Chemical oxidation involves using strong oxidising agents, such as persulfates or ozone, to degrade PFAS compounds. This method can break down PFAS into less harmful substances, but its effectiveness varies depending on the specific PFAS and the conditions of the treatment.

- **Reduction Technologies:** Chemical reduction processes use reducing agents to convert PFAS into less toxic compounds. These methods are still being researched and developed, with promising results for certain PFAS types. However, more research is needed to determine their effectiveness across a range of PFAS and environmental conditions.

C. Bioremediation:

- **Microbial Degradation:** Bioremediation involves using microorganisms to degrade PFAS. While some microorganisms can break down specific PFAS compounds, the process is generally slow and not yet effective for all types of PFAS. Research is ongoing to identify and optimise microbes capable of degrading PFAS more efficiently.

- **Enzymatic Approaches:** Enzymatic bioremediation uses

enzymes to break down PFAS compounds. While still in the experimental stage, certain enzymes have shown potential in degrading PFAS under controlled conditions. This approach could complement other remediation technologies in the future.

3. In-Situ and Ex-Situ Remediation
A. In-Situ Remediation:

- **Injection of Reactants:** In-situ remediation involves injecting chemical reactants, such as activated carbon or reactive materials, directly into contaminated groundwater or soil. This method aims to treat contaminants in place, reducing the need for excavation or pump-and-treat systems.
- **Effectiveness and Considerations:** In-situ remediation can be effective for treating PFAS contamination in groundwater and soil. The success of this approach depends on factors such as the depth of contamination, soil type, and the ability of the reactants to reach and treat the PFAS. Monitoring and managing the process are crucial to ensure effectiveness and avoid unintended consequences.

B. Ex-Situ Remediation:

- **Excavation and Treatment:** Ex-situ remediation involves excavating contaminated soil or sediments and treating them above ground using various methods, such as thermal desorption or chemical treatment. This approach can be effective for removing PFAS from soil and sediment but can be costly and disruptive.
- **Containment and Disposal:** For some PFAS-

contaminated materials, containment and disposal may be the most practical solution. This involves securely storing the contaminated materials in specialised facilities to prevent further environmental impact. The cost and logistics of containment and disposal need to be carefully managed.

4. Future Prospects and Innovations
A. Research and Development:

- **Innovative Materials and Technologies:** Ongoing research is focused on developing new materials and technologies for PFAS removal and destruction. This includes exploring novel adsorbents, reactive materials, and advanced oxidation processes. Innovations in these areas could enhance the effectiveness and efficiency of PFAS remediation.

- **Integration of Remediation Technologies:** Combining multiple remediation technologies in a treatment train or integrated system may offer improved results. For example, using a combination of adsorption, oxidation, and bioremediation could address a broader range of PFAS compounds and contamination scenarios.

B. Policy and Regulation:

- **Supporting Research and Implementation:** Policymakers and regulators play a crucial role in supporting research and the implementation of advanced remediation technologies. Funding for research, development of treatment standards, and incentives for adopting innovative solutions can drive progress in managing PFAS contamination.

- **Promoting Best Practices:** Establishing best practices and guidelines for PFAS remediation can help ensure the effective and responsible use of new technologies. Collaboration between industry, government, and research institutions is essential for advancing the state of PFAS remediation and addressing the challenges of contamination.

5. Conclusion

The analysis of current remediation techniques highlights the complexity of managing PFAS contamination and the need for continued innovation and research. While conventional methods like activated carbon filtration, ion exchange, and reverse osmosis remain effective, advanced technologies such as high-temperature incineration, chemical oxidation, and bioremediation offer promising solutions. Future advancements in detection and remediation technologies, combined with supportive policies and collaborative efforts, will be crucial for addressing the challenges posed by PFAS contamination and ensuring effective environmental protection.

Future Prospects for Improving PFAS Removal and Containment Methods

As PFAS (Per- and Polyfluoroalkyl Substances) continue to present challenges due to their persistence and widespread contamination, the need for advanced and innovative removal and containment methods is becoming increasingly critical. This section explores future prospects for improving PFAS removal and containment, focusing on emerging technologies, research developments, and potential strategies for more effective management of PFAS contamination.

1. Emerging Technologies in PFAS Removal

A. Advanced Oxidation Processes (AOPs):

- **Principles of AOPs:** Advanced Oxidation Processes (AOPs) use powerful oxidising agents and catalysts to break down persistent organic contaminants, including PFAS. Techniques such as Fenton's reagent, photo-Fenton, and UV/H2O2 (ultraviolet light combined with hydrogen peroxide) generate hydroxyl radicals that react with PFAS, leading to their degradation into less harmful substances.

- **Research and Development:** Research is focused on improving the efficiency of AOPs by optimising reaction conditions and exploring new catalyst materials. Recent advancements include the development of novel photocatalysts and the integration of AOPs with other treatment methods to enhance PFAS removal. The challenge remains to scale up these processes for practical, large-scale applications while managing costs and by-products.

B. Electrochemical Technologies:

- **Electrochemical Oxidation:** Electrochemical oxidation uses an electric current to drive oxidation reactions, breaking down PFAS compounds in water. Techniques such as electrochemical advanced oxidation processes (EAOPs) and electrochemical degradation have shown promise in laboratory settings for treating PFAS contamination.

- **Advantages and Challenges:** Electrochemical technologies offer the advantage of in-situ treatment with minimal chemical usage. However, scaling these technologies for large-scale applications requires further research to address challenges such as electrode fouling, efficiency, and cost-effectiveness.

C. Nanomaterials and Nanotechnology:

- **Nanomaterial-Based Adsorbents:** Nanomaterials, such as nanoscale zero-valent iron (nZVI) and carbon-based nanomaterials, offer high surface areas and reactivity for PFAS adsorption and degradation. These materials can be engineered to target specific PFAS compounds and enhance removal efficiency.
- **Innovations and Applications:** Ongoing research is exploring the use of nanomaterials in both water treatment and soil remediation. Developments include functionalised nanomaterials that can selectively adsorb PFAS or facilitate chemical reactions to break down these contaminants. Challenges include the potential environmental impact of nanomaterials and ensuring their safe and effective deployment.

2. Improved Containment Strategies
A. Barrier Technologies:

- **Permeable Reactive Barriers (PRBs):** Permeable reactive barriers are installed underground to intercept and treat contaminated groundwater as it flows through the barrier. Materials such as zero-valent iron, activated carbon, and organoclays are used to degrade or adsorb PFAS within the barrier.
- **Effectiveness and Considerations:** PRBs can be effective in reducing PFAS concentrations in groundwater over extended periods. Design considerations include the

choice of reactive materials, barrier placement, and potential maintenance needs. Research is focused on enhancing the longevity and efficiency of PRBs for PFAS treatment.

B. Phytoremediation:

- **Use of Plants for PFAS Removal:** Phytoremediation involves using plants to absorb, accumulate, or degrade contaminants from soil or water. Research is investigating plant species and genetic modifications that may enhance the uptake and transformation of PFAS compounds.
- **Potential and Limitations:** While phytoremediation offers a potentially sustainable approach to managing PFAS contamination, it is generally applicable to specific scenarios and contaminants. The effectiveness of phytoremediation for PFAS is still under investigation, and practical applications may be limited by factors such as plant growth conditions and the types of PFAS present.

3. Integration of Technologies
A. Hybrid Treatment Systems:

- **Combination of Methods:** Integrating multiple treatment technologies can enhance PFAS removal efficiency and address the limitations of individual methods. For example, combining activated carbon adsorption with advanced oxidation processes or electrochemical techniques can provide comprehensive treatment of PFAS-contaminated water.
- **Benefits and Challenges:** Hybrid systems can offer improved performance and flexibility in addressing a range

of PFAS compounds. However, designing and operating these systems require careful consideration of compatibility, cost, and operational complexity.

B. Modular and Scalable Solutions:

- **Modular Systems:** Modular treatment systems allow for flexibility and scalability in addressing PFAS contamination. These systems can be adapted to different contamination levels and site conditions, providing a versatile approach to remediation.
- **Scalability and Implementation:** Developing scalable solutions that can be efficiently deployed in various contexts is essential for addressing widespread PFAS contamination. Research and development efforts are focused on creating modular systems that are cost-effective, easy to implement, and capable of handling large volumes of contaminated water or soil.

4. Policy and Research Support
A. Funding and Incentives:

- **Support for Innovation:** Government and private sector funding are crucial for supporting research and development of new PFAS removal and containment technologies. Grants, incentives, and partnerships can drive innovation and facilitate the deployment of advanced solutions.
- **Collaborative Research:** Collaborative research efforts involving academia, industry, and government agencies can accelerate the development and adoption of new technologies. Sharing knowledge, resources, and expertise

helps to address the complex challenges associated with PFAS contamination.

B. Regulatory and Policy Framework:

- **Standards and Guidelines:** Developing and updating regulatory standards and guidelines for PFAS treatment can drive the adoption of effective technologies. Clear guidelines on treatment performance, monitoring, and reporting help ensure that remediation efforts meet regulatory requirements.
- **Encouraging Best Practices:** Promoting best practices and standards for PFAS removal and containment supports the implementation of effective technologies. Industry organisations, regulatory bodies, and research institutions play a role in setting and enforcing standards that guide remediation efforts.

5. Conclusion

The future of PFAS removal and containment lies in the continued advancement and integration of innovative technologies. Emerging methods, such as advanced oxidation processes, electrochemical technologies, and nanomaterials, hold promise for more effective treatment of PFAS contamination. Improved containment strategies, including barrier technologies and phytoremediation, offer additional options for managing PFAS. Collaboration between researchers, policymakers, and industry stakeholders is essential for advancing these technologies and developing scalable, cost-effective solutions. As research progresses and new technologies emerge, they will play a crucial role in addressing the challenges of PFAS contamination and ensuring the protection of public health and the environment.

Future Prospects for Improving PFAS Removal and Containment Methods

As the challenges associated with PFAS (Per- and Polyfluoroalkyl Substances) contamination continue to evolve, it is crucial to explore future prospects for enhancing removal and containment methods. The persistence and widespread nature of PFAS necessitate ongoing innovation in remediation technologies and strategies. This section delves into the potential advancements and research directions that could significantly improve PFAS management, offering insights into both emerging technologies and strategic approaches for effective treatment and containment.

1. Innovative Approaches to PFAS Destruction

A. Advanced Photocatalysis:

- **Mechanism and Advancements:** Photocatalysis involves using light-activated catalysts to facilitate chemical reactions that break down contaminants. Recent advancements in photocatalytic materials, such as titanium dioxide (TiO_2) and graphitic carbon nitride (g-C_3N_4), have shown potential for degrading PFAS under UV or visible light. Researchers are developing more efficient photocatalysts that can operate under ambient conditions and with lower energy requirements.

- **Research and Applications:** Ongoing research is focused on enhancing the efficiency and stability of photocatalysts for PFAS degradation. Innovations include the development of hybrid photocatalysts that combine multiple materials to improve performance and durability. These advancements could lead to scalable, in-situ treatment solutions for PFAS-contaminated water and soil.

B. Chemical Looping Technologies:

- **Concept and Implementation:** Chemical looping technologies involve the use of metal oxides or other materials in a cyclic process to oxidise and destroy PFAS compounds. This method offers the potential for efficient and complete destruction of PFAS without generating significant by-products.
- **Development and Challenges:** Chemical looping is still in the experimental phase for PFAS treatment. Researchers are working to optimise the reaction conditions, material selection, and process design to achieve effective and sustainable PFAS destruction. Overcoming challenges such as material costs and process scalability will be key to practical implementation.

C. Electrocatalytic Degradation:

- **Principle and Progress:** Electrocatalytic degradation uses electrical energy to drive chemical reactions that break down PFAS compounds. This approach can be applied in electrochemical cells or reactors, offering the potential for targeted and efficient treatment. Recent developments focus on improving the electrocatalysts used in these processes to enhance degradation rates and efficiency.
- **Future Research Directions:** Research is exploring various electrode materials and reaction conditions to optimise electrocatalytic degradation of PFAS. Innovations in electrode design and reactor configuration could improve the effectiveness and practicality of this technology for PFAS remediation.

2. Integrating Remediation Technologies
A. Multi-Stage Treatment Systems:

- **Concept and Advantages:** Multi-stage treatment systems combine different remediation technologies to address the complex nature of PFAS contamination. For example, a system might integrate activated carbon adsorption, advanced oxidation, and bioremediation to provide comprehensive treatment. This approach can target a broader range of PFAS compounds and enhance overall removal efficiency.

- **Implementation and Optimisation:** Designing and implementing multi-stage systems require careful consideration of compatibility, process control, and operational costs. Research and pilot projects are exploring optimal configurations and combinations of technologies to maximise treatment effectiveness while managing costs and resource use.

B. Modular and Adaptable Systems:

- **Benefits and Flexibility:** Modular treatment systems offer flexibility and scalability, allowing for adjustments based on contamination levels and site conditions. These systems can be customised and expanded as needed, providing adaptable solutions for various PFAS contamination scenarios.

- **Future Developments:** Advances in modular design and technology integration are expected to enhance the functionality and efficiency of these systems. Innovations may include improved materials, automated controls, and remote monitoring capabilities to optimise performance and streamline operations.

3. Advancements in Monitoring and Detection

A. Real-Time Monitoring Technologies:

- **Innovations in Sensors:** Real-time monitoring technologies, such as advanced sensors and analytical instruments, are being developed to provide continuous, accurate measurements of PFAS concentrations in water and soil. These technologies enable prompt detection of contamination changes and more effective management of treatment systems.

- **Applications and Benefits:** Real-time monitoring can improve the efficiency of remediation efforts by providing timely data for decision-making. Innovations include portable sensors for field use and remote sensing technologies for large-scale monitoring. Enhanced monitoring capabilities will support better management and reduction of PFAS contamination.

B. High-Resolution Mass Spectrometry:

- **Advanced Analytical Techniques:** High-resolution mass spectrometry (HRMS) is a powerful analytical technique used to identify and quantify trace levels of PFAS compounds. Recent advancements in HRMS have improved the sensitivity and specificity of PFAS analysis, allowing for more comprehensive detection of individual PFAS species.

- **Future Prospects:** Ongoing research is focused on further enhancing HRMS capabilities to detect lower concentrations of PFAS and analyse complex samples. Improvements in sample preparation and data analysis techniques will contribute to more accurate and detailed monitoring of PFAS contamination.

4. Policy and Collaborative Efforts

A. Supporting Research and Development:

- **Funding and Incentives:** Continued investment in research and development is crucial for advancing PFAS remediation technologies. Governments, private sector organisations, and research institutions should collaborate to provide funding, resources, and incentives for innovation.

- **Collaborative Research Initiatives:** Collaborative research efforts involving multiple stakeholders can accelerate the development and implementation of new technologies. Partnerships between academia, industry, and government agencies will drive progress and facilitate the adoption of effective solutions.

B. Regulatory and Policy Frameworks:

- **Updating Regulations:** Developing and updating regulatory frameworks to reflect advancements in PFAS treatment technologies will support effective management and enforcement. Clear guidelines and standards for PFAS removal and containment will help ensure that new technologies are applied effectively and safely.

- **Promoting Best Practices:** Encouraging the adoption of best practices and standards for PFAS remediation can drive improvements in technology and operational efficiency. Industry organisations and regulatory bodies play a key role in setting and promoting these practices.

5. Conclusion

The future of PFAS removal and containment lies in the continued development and integration of innovative technologies and strategies. Advancements in methods such as advanced photocatalysis, chemical looping, and electrocatalytic degradation hold promise for more effective PFAS treatment. Integrating various technologies into multi-stage and modular systems can enhance overall remediation efforts. Improved monitoring and detection technologies will support more precise and responsive management of PFAS contamination. Collaborative research, investment, and supportive regulatory frameworks are essential for driving progress and achieving successful outcomes in PFAS remediation. As the field continues to evolve, these advancements will play a critical role in addressing the challenges posed by PFAS contamination and protecting public health and the environment.

Public Awareness and Advocacy

Exploring Public Awareness Regarding PFAS Contamination and Its Health Impacts

The issue of PFAS (Per- and Polyfluoroalkyl Substances) contamination has gradually entered the public consciousness, driven by increasing evidence of its severe environmental and health impacts. This section delves into the current state of public awareness regarding PFAS, examining how knowledge about these hazardous substances has evolved and the factors influencing public understanding.

1. Evolution of Public Awareness

A. Historical Context:

- **Early Awareness:** Public awareness of PFAS began to rise in the early 2000s, primarily due to scientific studies highlighting the widespread environmental contamination and potential health risks associated with these substances. Initial concerns were largely focused on specific incidents, such as contamination near military bases and industrial facilities using PFAS-containing firefighting foams.

- **Scientific Breakthroughs:** Research advancements over the past two decades have significantly expanded the understanding of PFAS. Studies linking PFAS exposure to various health issues, including cancer, liver damage, and immune system effects, have heightened public concern. The release of comprehensive studies and reports from organisations like the Environmental Protection Agency (EPA) and the Agency for Toxic Substances and Disease Registry (ATSDR) has played a crucial role in educating the public.

B. Current State of Knowledge:

- **General Awareness:** Today, there is a growing awareness of PFAS among the general public, but knowledge varies significantly across different regions and demographics. Media coverage, community outreach programmes, and educational campaigns have contributed to increased awareness, though many people remain unfamiliar with the specifics of PFAS contamination and its long-term effects.

- **Misconceptions and Gaps:** Despite progress, misconceptions about PFAS persist. Some individuals may not fully grasp the extent of PFAS contamination, its sources, or the severity of its health impacts. Gaps in public understanding can hinder effective advocacy and action, underscoring the need for ongoing education and accurate information dissemination.

2. Factors Influencing Public Awareness
A. Media Coverage:

- **Role of the Media:** The media has played a pivotal role in raising awareness about PFAS contamination. Investigative journalism and news reports have brought attention to major contamination incidents, highlighting the associated health risks and regulatory shortcomings. High-profile cases and exposés often generate widespread public interest and concern.

- **Challenges:** Media coverage can sometimes be sensationalised, focusing on specific incidents without providing a comprehensive overview of the broader issue. Ensuring balanced and accurate reporting is essential to foster a well-informed public. Additionally, the complexity of PFAS-related science and regulations can make it challenging for media outlets to convey detailed information effectively.

B. Environmental NGOs and Advocacy Groups:

- **Influence of NGOs:** Environmental non-governmental organisations (NGOs) and advocacy groups have been instrumental in raising awareness about PFAS. These organisations often conduct research, publish reports, and engage in public education campaigns. Their efforts help to bridge the gap between scientific knowledge and public understanding.
- **Advocacy and Campaigns:** NGOs frequently lead advocacy campaigns to highlight PFAS issues and push for regulatory changes. These campaigns can include public demonstrations, petitions, and lobbying efforts aimed at government officials and policymakers. The work of these groups is crucial in driving public engagement and influencing policy decisions.

C. Community Engagement:

- **Grassroots Efforts:** Community groups and local activists play a significant role in raising awareness about PFAS contamination at the grassroots level. These efforts often stem from direct experiences of contamination and health impacts, prompting local residents to seek information and advocate for change.
- **Case Studies:** Community-driven initiatives have led to increased scrutiny of local contamination issues and have sometimes resulted in tangible policy changes. For

example, communities affected by PFAS contamination near military bases or industrial facilities have successfully mobilised to demand clean-up efforts and better regulatory oversight.

3. Case Studies of Public Awareness Efforts
A. The "Toxic PFAS" Campaign in Scotland:

- **Background and Impact:** In Scotland, a grassroots campaign known as "Toxic PFAS" emerged in response to local contamination concerns. Residents discovered PFAS contamination in drinking water sources and sought to raise awareness about the issue. The campaign utilised social media, local meetings, and public protests to engage the community and press for action.

- **Outcomes:** The campaign succeeded in drawing national attention to the issue, leading to increased media coverage and government investigations. As a result, regulatory bodies implemented stricter monitoring and response measures, and public awareness of PFAS issues in Scotland was significantly heightened.

B. The "Fight for Clean Water" Movement in the US:

- **Background and Achievements:** In the United States, the "Fight for Clean Water" movement has been instrumental in addressing PFAS contamination. This movement has brought together environmental groups, affected communities, and health advocates to push for regulatory changes and remediation efforts. High-profile cases, such as the contamination in Flint, Michigan, and various military bases, have driven public and governmental responses.

- **Impact on Policy:** The movement has led to significant policy changes, including new regulations on PFAS levels in drinking water and increased funding for research and remediation. The advocacy efforts have also inspired similar movements in other countries, contributing to a global push for PFAS action.

4. Challenges and Opportunities for Increasing Awareness
A. Addressing Knowledge Gaps:

- **Educational Initiatives:** To bridge knowledge gaps, it is essential to invest in educational initiatives that provide accurate, accessible information about PFAS. Public health campaigns, educational materials, and community workshops can help enhance understanding and address misconceptions.
- **Collaboration with Experts:** Collaborating with scientists, health professionals, and environmental experts can ensure that public information is accurate and up-to-date. Engaging experts in outreach efforts can also help to clarify complex scientific concepts and improve public comprehension.

B. Enhancing Communication Strategies:

- **Effective Messaging:** Crafting clear and compelling messages about PFAS contamination and health risks is crucial for engaging the public. Tailoring communication strategies to different audiences and using multiple channels, such as social media, traditional media, and community events, can maximise outreach efforts.

- **Utilising Technology:** Leveraging technology, such as mobile apps and online platforms, can facilitate real-time information sharing and community engagement. Technology can also support monitoring and reporting of contamination, providing valuable tools for advocacy and public awareness.

5. Conclusion

Public awareness of PFAS contamination and its health impacts has evolved significantly over recent years, driven by scientific research, media coverage, and advocacy efforts. While awareness has increased, there remain gaps in knowledge and challenges in effectively communicating the complexities of PFAS issues. Continued efforts by environmental NGOs, media, community groups, and public health organisations are crucial for enhancing understanding, driving policy change, and fostering informed public action. As awareness grows and advocacy efforts intensify, addressing PFAS contamination and its associated risks will become increasingly feasible, paving the way for improved environmental and public health outcomes.

Case Studies of Successful Advocacy Efforts and Their Impact on Policy Changes

Successful advocacy efforts have played a pivotal role in addressing PFAS (Per- and Polyfluoroalkyl Substances) contamination, influencing policy changes, and raising public awareness. This section explores several notable case studies where advocacy groups, media, and community action have led to significant outcomes in the fight against PFAS pollution. By examining these examples, we can better understand the strategies that have proven effective and the impact they have had on both local and national levels.

1. The Hoosick Falls Contamination Crisis

A. Background and Discovery:

- **The Incident:** In Hoosick Falls, New York, residents began to notice health issues and a decline in water quality around 2014. Investigations revealed that PFAS, particularly PFOA (Perfluorooctanoic Acid), had contaminated the town's drinking water supply. The contamination was traced back to a nearby manufacturing plant that used PFAS in its products.
- **Community Response:** The local community, led by concerned residents and local activists, quickly mobilised to address the issue. The group, including individuals like Michael Hickey, played a crucial role in raising awareness and pushing for investigations into the contamination.

B. Advocacy and Media Involvement:

- **Media Coverage:** The story gained national attention thanks to investigative journalism by outlets such as the Times Union and the New York Times. Media coverage brought the issue to the forefront, exposing the extent of the contamination and the potential health risks associated with PFAS exposure.
- **Legal and Regulatory Actions:** The heightened media scrutiny and public pressure led to legal action against the responsible companies. The state of New York and the Environmental Protection Agency (EPA) initiated investigations and enforcement actions. In response to the crisis, New York State implemented stricter PFAS regulations and provided funding for water treatment and health studies.

C. Impact and Legacy:

- **Policy Changes:** The Hoosick Falls case prompted broader discussions about PFAS contamination, leading to legislative and regulatory changes. New York State introduced new limits on PFAS in drinking water and expanded testing requirements. The case also contributed to increased awareness and research into PFAS contamination nationwide.
- **Community Empowerment:** The successful advocacy efforts in Hoosick Falls empowered other communities facing similar issues to take action. The case highlighted the importance of community engagement and media involvement in addressing environmental health crises.

2. The "Toxic PFAS" Campaign in Scotland
A. Background and Campaign Initiation:

- **Discovery of Contamination:** In Scotland, PFAS contamination was identified in various regions, including sites associated with firefighting foams and industrial activities. Concerns about the impact on drinking water and public health prompted local residents and environmental groups to take action.
- **Campaign Launch:** The "Toxic PFAS" campaign was launched to address the contamination and push for stronger regulations. The campaign involved a coalition of environmental organisations, local activists, and concerned citizens working together to raise awareness and advocate for change.

B. Strategies and Achievements:

- **Public Engagement:** The campaign employed a range of strategies, including social media outreach, public meetings, and protests. These efforts were aimed at educating the public about PFAS risks and mobilising support for regulatory changes.
- **Regulatory Impact:** The campaign successfully influenced policy decisions at both local and national levels. Scottish authorities implemented stricter monitoring and response measures for PFAS contamination, and new regulations were introduced to limit the use of PFAS-containing products.

C. Broader Implications:

- **National and International Influence:** The success of the "Toxic PFAS" campaign in Scotland set a precedent for other regions facing PFAS contamination. The campaign's achievements contributed to increased global awareness of PFAS issues and inspired similar advocacy efforts in other countries.
- **Continued Advocacy:** The campaign continues to advocate for further measures to address PFAS contamination, including expanded research and improved cleanup efforts. The ongoing efforts highlight the need for sustained advocacy and public engagement in addressing environmental health issues.

3. The "Fight for Clean Water" Movement in the United States

A. Movement Overview:

- **Background:** The "Fight for Clean Water" movement

emerged in response to widespread PFAS contamination across the United States. This movement brought together environmental organisations, community groups, and affected individuals to advocate for stronger regulations and remedial action.

- **Key Actions:** The movement has involved a range of activities, including public demonstrations, legal challenges, and lobbying efforts. High-profile cases of PFAS contamination, such as those affecting communities near military bases and industrial sites, have been central to the movement's advocacy.

B. Major Achievements:

- **Legislative Successes:** The movement has achieved notable successes in advancing PFAS regulation. For example, the introduction of the PFAS Action Act in Congress aims to establish federal standards for PFAS in drinking water and increase funding for research and remediation. State-level initiatives have also led to new regulations and clean-up efforts.
- **Public Awareness:** The movement has significantly raised public awareness about PFAS contamination. Media coverage, combined with advocacy campaigns, has educated the public about the risks of PFAS and the need for regulatory action. The movement's efforts have prompted increased scrutiny of PFAS-related practices and led to more robust regulatory frameworks.

C. Ongoing Challenges and Future Directions:

- **Sustaining Momentum:** While the movement has

achieved significant progress, challenges remain, including the need for continued advocacy and funding. Ensuring that regulatory changes are implemented effectively and that communities receive the support they need is essential for sustaining momentum.

- **Expanding Scope:** The movement's success has highlighted the need to address PFAS contamination on a global scale. Continued international collaboration and advocacy are necessary to address PFAS issues comprehensively and protect public health worldwide.

4. The Role of Grassroots Movements in Policy Change
A. Empowering Local Communities:

- **Grassroots Advocacy:** Grassroots movements play a crucial role in raising awareness and driving policy changes related to PFAS contamination. Local communities often initiate advocacy efforts based on direct experiences with contamination, leading to targeted and effective campaigns.
- **Success Stories:** Examples of successful grassroots advocacy include communities near contaminated sites that have mobilised to demand clean-up and regulatory changes. These efforts demonstrate the power of local action in influencing policy and achieving meaningful results.

B. Collaborative Efforts:

- **Partnerships and Alliances:** Collaborative efforts between grassroots movements, environmental organisations, and policymakers can enhance the effectiveness of advocacy campaigns. Building alliances and fostering cooperation can lead to more comprehensive solutions and increased impact.

- **Lessons Learned:** Successful grassroots campaigns provide valuable lessons for future advocacy efforts. Strategies such as building strong community networks, leveraging media coverage, and engaging with policymakers can help maximise the effectiveness of advocacy campaigns.

5. Conclusion

Case studies of successful advocacy efforts reveal the significant impact that environmental NGOs, media, and community groups can have on addressing PFAS contamination and influencing policy changes. From local grassroots movements to national campaigns, these efforts have driven increased awareness, regulatory action, and public engagement. By examining these examples, we can identify effective strategies and approaches for future advocacy, ensuring continued progress in the fight against PFAS pollution and its associated risks.

Case Studies of Successful Advocacy Efforts and Their Impact on Policy Changes

Advocacy efforts aimed at addressing PFAS (Per- and Polyfluoroalkyl Substances) contamination have led to significant policy changes and increased public awareness. This section delves into several detailed case studies that showcase the strategies employed by various groups and the outcomes of their efforts. These examples illustrate how targeted advocacy can effect meaningful change and highlight lessons learned for future initiatives.

1. The Michigan PFAS Crisis

A. Background and Early Efforts:

- **Initial Discovery:** In Michigan, PFAS contamination was discovered in several communities, notably in areas surrounding military bases and industrial sites. The contamination was linked to the use of PFAS-containing firefighting foams and industrial processes. As reports emerged linking PFAS to health problems, residents and local officials began to demand action.

- **Community Mobilisation:** Local communities, such as those in the city of Flint, organised grassroots efforts to address the contamination. These groups formed coalitions and worked with environmental organisations to raise awareness about the health risks associated with PFAS. They also sought assistance from legal and environmental experts to advocate for remediation and regulatory changes.

B. Key Strategies and Achievements:

- **Public Awareness Campaigns:** Advocates utilised various platforms to raise public awareness, including social media campaigns, community meetings, and public demonstrations. High-profile events and media coverage played a crucial role in keeping PFAS contamination in the public eye and driving action.
- **Legislative Action:** Advocacy efforts led to the introduction of state legislation aimed at addressing PFAS contamination. Michigan passed laws to regulate PFAS in drinking water and required monitoring and reporting of PFAS levels. The state's efforts also included funding for remediation projects and health studies.
- **Impact:** The Michigan PFAS crisis brought national attention to PFAS contamination and highlighted the

need for comprehensive regulations. The state's response served as a model for other regions facing similar issues, prompting legislative action and increased funding for PFAS-related research and clean-up efforts across the United States.

2. The European Union's Approach to PFAS Regulation
A. Regulatory Framework and Initiatives:

- **Initial Steps:** The European Union (EU) has taken a proactive approach to PFAS regulation, recognising the potential risks posed by these substances. The EU's approach includes a range of measures, from restricting the use of specific PFAS chemicals to promoting research and innovation in PFAS alternatives.
- **EU-wide Initiatives:** The European Chemicals Agency (ECHA) has been instrumental in advancing PFAS regulation. ECHA has proposed restrictions on certain PFAS chemicals and initiated assessments of their environmental and health impacts. Additionally, the EU's Green Deal aims to address the broader issue of chemical pollution, including PFAS.

B. Key Strategies and Achievements:

- **Public Consultations and Research:** The EU has conducted public consultations and research to inform its regulatory approach. By engaging with stakeholders, including environmental groups, industry representatives, and researchers, the EU has developed a comprehensive strategy for managing PFAS risks.
- **Legislative Measures:** The EU has introduced regulations

such as the REACH (Registration, Evaluation, Authorisation and Restriction of Chemicals) framework, which includes provisions for assessing and managing PFAS chemicals. The EU's approach has set a high standard for chemical regulation and has influenced global discussions on PFAS.

- **Impact:** The EU's regulatory measures have led to increased awareness of PFAS risks and prompted other countries to consider similar approaches. The EU's proactive stance has contributed to the development of safer alternatives and has driven international collaboration on PFAS issues.

3. The "Safe Drinking Water" Campaign in Australia
A. Campaign Overview:

- **Discovery of Contamination:** In Australia, PFAS contamination was identified in several regions, particularly around military bases and industrial sites. The contamination led to widespread concerns about the safety of drinking water and potential health impacts.
- **Campaign Launch:** The "Safe Drinking Water" campaign was launched by a coalition of environmental organisations, community groups, and affected residents. The campaign aimed to raise awareness about PFAS contamination, advocate for regulatory changes, and secure funding for remediation and health studies.

B. Strategies and Achievements:

- **Advocacy and Public Engagement:** The campaign employed a range of advocacy strategies, including public demonstrations, media campaigns, and lobbying efforts. High-profile cases and personal stories helped to highlight the severity of the contamination and the need for action.

- **Regulatory Impact:** The campaign's efforts led to increased government action on PFAS contamination. Australian authorities implemented new regulations for PFAS in drinking water, funded research on health impacts, and launched clean-up projects in affected areas.
- **Impact:** The "Safe Drinking Water" campaign contributed to significant policy changes and raised public awareness about PFAS contamination. The campaign's success also demonstrated the effectiveness of collaborative advocacy efforts and underscored the importance of community involvement in addressing environmental issues.

4. The Role of Social Media in Advocacy
A. Amplifying Voices and Raising Awareness:

- **Social Media Platforms:** Social media has become a powerful tool for raising awareness about PFAS contamination. Platforms such as Twitter, Facebook, and Instagram allow advocates to share information, mobilise support, and engage with a global audience.
- **Successful Campaigns:** Several advocacy campaigns have effectively used social media to amplify their messages. For example, the hashtag #PFASFree and similar campaigns have helped to spread information about PFAS risks and advocate for regulatory changes.

B. Challenges and Opportunities:

- **Information Overload:** While social media offers

opportunities for outreach, it also presents challenges such as information overload and misinformation. Ensuring that accurate and reliable information is shared is crucial for effective advocacy.

- **Engagement and Action:** Social media campaigns can drive real-world action by encouraging followers to participate in petitions, attend events, and contact policymakers. Leveraging social media effectively requires strategic planning and ongoing engagement with the audience.

5. Conclusion

Case studies of successful advocacy efforts reveal the significant impact that targeted campaigns, community action, and media involvement can have on addressing PFAS contamination and driving policy changes. From local grassroots movements to international initiatives, these efforts have led to increased awareness, regulatory advancements, and improved public health protections. By learning from these examples and continuing to advocate for change, we can build on past successes and advance the fight against PFAS pollution, ensuring a safer and healthier environment for all.

Impact on Local Communities – Case Studies

Detailed Accounts of Communities in the UK Directly Affected by PFAS Contamination

PFAS (Per- and Polyfluoroalkyl Substances) contamination has significantly impacted various communities across the UK. This section presents detailed accounts of several affected areas, focusing on the experiences of residents and the local response to the contamination. By examining these case studies, we can gain a clearer understanding of the far-reaching effects of PFAS pollution on communities and the challenges they face.

1. The Case of Larkhill, Wiltshire

A. Discovery and Initial Response:

- **Contamination Source:** In Larkhill, a military training area in Wiltshire, PFAS contamination was discovered in 2015. The contamination was linked to the use of firefighting foams containing PFAS, which had been used extensively on the site over several decades. The detection of elevated PFAS levels in local groundwater and drinking water sources led to immediate concerns among residents.

- **Community Reaction:** Residents of Larkhill were alarmed when they learned about the contamination. The discovery prompted a wave of community meetings, protests, and appeals for action. Local leaders, including the parish council and concerned citizens, spearheaded efforts to address the issue, demanding that the government and the Ministry of Defence take responsibility for the contamination and provide solutions.

B. Impact on Residents:

- **Health Concerns:** Many residents expressed concerns about potential health impacts related to PFAS exposure. Common symptoms reported included gastrointestinal issues, skin conditions, and concerns about long-term health effects such as cancer. The uncertainty surrounding health risks added to the community's anxiety.
- **Economic Disruption:** The contamination affected property values and local businesses. Prospective homebuyers were deterred by the contamination, leading to a decline in property values. Local businesses, particularly those relying on clean water for their operations, faced challenges in maintaining their services and customer trust.

C. Response and Remediation Efforts:

- **Government and Military Actions:** The Ministry of Defence acknowledged the contamination and committed to funding remediation efforts. Measures included cleaning up contaminated sites, providing alternative water supplies, and conducting health studies to assess the impact on residents.
- **Community Involvement:** Local residents and organisations actively participated in the response process, advocating for transparency and effective solutions. The community's involvement helped ensure that their concerns were addressed and that the remediation efforts were carried out with their best interests in mind.

2. The Situation in Fife, Scotland

A. Contamination and Discovery:

- **Source of Contamination:** In Fife, Scotland, PFAS contamination was identified around former military airfields where firefighting foams had been used. The contamination affected local water supplies and raised concerns about environmental and public health.
- **Community Response:** Residents in affected areas, such as those near the former RAF bases, organised meetings and campaigns to demand action from local authorities and the Scottish Government. Their efforts included petitions, public demonstrations, and media campaigns to raise awareness about the contamination.

B. Effects on Local Population:

- **Health Impact:** The community expressed concerns about the potential health impacts of PFAS exposure. There were reports of increased rates of certain health conditions among residents, which they attributed to the contamination. These concerns were exacerbated by the lack of comprehensive health studies at the time.
- **Economic Impact:** The contamination had significant economic repercussions for the affected areas. Local farmers and businesses faced challenges due to the contamination of soil and water, impacting agricultural productivity and local economies. Additionally, the contamination led to increased costs for water treatment and environmental monitoring.

C. Government and Community Actions:

- **Regulatory Measures:** The Scottish Government introduced new regulations and funding for PFAS remediation in response to the contamination. Measures included enhanced monitoring of water supplies and support for affected communities.
- **Community Advocacy:** Community groups played a crucial role in advocating for action and holding authorities accountable. Their persistent efforts helped to drive policy changes and ensure that the contamination was addressed in a timely manner.

3. The Impact on Dumfries and Galloway
A. Contamination Discovery:

- **Source and Discovery:** In Dumfries and Galloway, PFAS contamination was linked to the use of firefighting foams at local military training grounds. Elevated PFAS levels were detected in nearby water sources, prompting an investigation into the extent of the contamination.
- **Community Reaction:** The discovery of contamination led to widespread concern among residents. Community groups and local authorities collaborated to address the issue, focusing on identifying the source of contamination and mitigating its effects on local water supplies.

B. Health and Economic Effects:

- **Health Issues:** Residents reported concerns about potential health risks, including the possibility of long-term exposure-related health conditions. The community sought assurances from health authorities and demanded comprehensive health studies to assess the impact of PFAS contamination.

- **Economic Consequences:** The contamination affected local businesses and agriculture, particularly those reliant on clean water for their operations. The community faced increased costs associated with water treatment and environmental remediation, impacting local economic stability.

C. Response and Solutions:

- **Local and National Efforts:** Local authorities worked with national agencies to address the contamination and implement remediation measures. The efforts included improving water treatment facilities, conducting environmental assessments, and providing support to affected residents.
- **Community Engagement:** The involvement of local residents and community groups was crucial in ensuring that the response was effective and responsive to their needs. Their advocacy helped to secure funding for remediation and promote transparency throughout the process.

4. Broader Socio-Economic Impacts
A. Property Values and Economic Activity:

- **Property Values:** PFAS contamination has had a direct impact on property values in affected areas. The presence of contamination often leads to decreased property values and challenges in selling homes, as prospective buyers are deterred by the potential health risks.
- **Economic Disruption:** Local economies in contaminated

areas face disruptions due to the costs of remediation and the impact on businesses reliant on clean water. Agricultural and industrial operations may experience reduced productivity and increased costs, affecting overall economic stability.

B. Social and Psychological Effects:

- **Community Well-being:** The uncertainty and stress associated with PFAS contamination can have significant social and psychological effects on residents. The fear of potential health impacts and the disruption of daily life can contribute to increased anxiety and reduced quality of life.
- **Social Cohesion:** Efforts to address PFAS contamination often lead to increased community engagement and activism. While this can strengthen social cohesion and foster a sense of solidarity, it can also lead to tensions between affected communities and authorities.

5. Conclusion

The impact of PFAS contamination on local communities in the UK has been profound, affecting health, economic stability, and social well-being. Through detailed case studies, we see the diverse experiences of residents and the responses of local and national authorities to the contamination. These accounts highlight the challenges faced by affected communities and underscore the importance of effective remediation, robust regulations, and ongoing support for those impacted by PFAS pollution. Understanding these case studies provides valuable insights into the broader implications of PFAS contamination and the critical need for comprehensive solutions to address this pressing environmental issue.

Personal Stories of Residents and Local Leaders Working to Address the Issues

In the face of PFAS (Per- and Polyfluoroalkyl Substances) contamination, many residents and local leaders have become champions of their communities, working tirelessly to address the issues and advocate for solutions. This section explores personal stories from individuals directly affected by PFAS contamination, shedding light on their experiences, challenges, and the efforts they have made to combat the problem.

1. The Story of Emma Williams in Larkhill, Wiltshire

A. Personal Experience:

Emma Williams, a lifelong resident of Larkhill, has been at the forefront of the community's response to PFAS contamination. As a mother of two young children, Emma's concerns about the health risks associated with contaminated water were immediate and deeply personal.

- **Initial Realisation:** Emma's journey began when she received a notification from the local council about elevated PFAS levels in the water supply. Her initial reaction was one of shock and fear, particularly given the recent birth of her second child. The uncertainty surrounding the potential health effects led her to seek more information and become involved in community efforts.

- **Advocacy and Activism:** Determined to protect her family and her community, Emma joined forces with local advocacy groups. She became a vocal advocate for transparency and action, organising community meetings, and collaborating with environmental organisations to raise awareness. Her efforts included drafting petitions, speaking at public forums, and engaging with the media to keep the issue in the public eye.

B. Challenges and Triumphs:

- **Health Concerns:** Emma faced significant challenges in her advocacy work, particularly regarding the lack of immediate and clear information about health risks. The ambiguity around potential health effects made it difficult to reassure fellow residents and push for specific health interventions.
- **Community Impact:** Despite these challenges, Emma's activism played a crucial role in galvanising the community and prompting local authorities to take action. Her persistence contributed to the allocation of funds for water treatment and the establishment of a health study to assess the impact of PFAS on residents.

2. The Efforts of Councillor John McGregor in Fife, Scotland
A. Local Leadership:

Councillor John McGregor, representing Fife, has been instrumental in addressing the PFAS contamination issue in his constituency. His role as a local leader gave him a unique platform to advocate for affected communities and coordinate response efforts.

- **Community Response:** Upon learning about PFAS contamination in the area, Councillor McGregor took immediate action to mobilise resources and support. He organised emergency meetings with local authorities, health officials, and environmental experts to develop a comprehensive response plan. His leadership ensured that the issue received the attention it warranted from both local and national bodies.

- **Public Engagement:** Councillor McGregor actively engaged with residents, hosting public forums to provide updates and answer questions. His approach emphasised transparency and community involvement, fostering a sense of trust and collaboration between authorities and affected residents.

B. Achievements and Obstacles:

- **Regulatory Changes:** Thanks to Councillor McGregor's efforts, the local council was able to push for stricter regulations on PFAS in drinking water. His advocacy led to increased funding for remediation projects and support for affected businesses and farms.
- **Ongoing Challenges:** Despite these successes, McGregor faced challenges related to bureaucratic delays and limited resources. Ensuring that remediation efforts were carried out effectively and addressing the health concerns of residents remained ongoing tasks.

3. The Advocacy of Sarah Johnson in Dumfries and Galloway
A. Grassroots Activism:

Sarah Johnson, a local business owner in Dumfries and Galloway, became a prominent advocate for her community following the discovery of PFAS contamination. Her personal stake in the issue was driven by her reliance on clean water for her agricultural business and concerns about the health of her family.

- **Community Campaigns:** Sarah launched a grassroots campaign to raise awareness about the contamination and push for action. Her campaign included organising community events, creating informational materials, and working with local media to highlight the issue. Her efforts were crucial in rallying support from both residents and local leaders.

- **Health and Economic Impact:** Sarah's campaign highlighted the dual impact of PFAS contamination on health and the local economy. She emphasised the need for immediate action to address contamination and support affected businesses, particularly those reliant on agricultural products.

B. Results and Continued Efforts:

- **Local and National Support:** Sarah's advocacy resulted in increased attention to the PFAS issue, leading to support from local government and national agencies. Her efforts contributed to the allocation of funds for water treatment and health studies, as well as the development of new regulations.
- **Ongoing Advocacy:** Despite these successes, Sarah continued to advocate for her community, focusing on ensuring that remediation efforts were effective and that affected residents received the support they needed. Her commitment to the cause underscored the importance of sustained advocacy in addressing environmental issues.

4. The Role of Community Groups and NGOs
A. Collaborative Efforts:

In addition to individual activists, various community groups and non-governmental organisations (NGOs) have played a significant role in addressing PFAS contamination. These organisations often work in partnership with affected communities, providing resources, expertise, and advocacy.

- **Environmental NGOs:** Organisations such as Friends of the Earth and the Environmental Defence Fund have been instrumental in raising awareness about PFAS contamination and advocating for policy changes. Their work includes conducting research, supporting local campaigns, and engaging with policymakers to drive regulatory reforms.
- **Local Support Groups:** Community-based support groups have provided crucial assistance to residents affected by PFAS contamination. These groups offer resources such as information about health risks, legal advice, and support for accessing remediation and compensation.

B. Impact and Challenges:

- **Successful Campaigns:** Collaborative efforts between community groups and NGOs have led to significant achievements, including regulatory changes, increased funding for remediation, and improved public awareness. These efforts highlight the power of collective action in addressing environmental issues.
- **Challenges Faced:** Despite their successes, these organisations face challenges such as limited funding, bureaucratic obstacles, and the need to navigate complex regulatory frameworks. Ensuring that their efforts lead to meaningful and lasting change requires continued dedication and resilience.

5. Conclusion

Personal stories from residents and local leaders affected by PFAS contamination illustrate the profound impact of environmental pollution on individuals and communities. Through their advocacy, resilience, and dedication, these individuals have played a crucial role in addressing the challenges posed by PFAS contamination. Their experiences underscore the importance of community involvement, effective leadership, and collaborative efforts in driving meaningful change and ensuring a safer environment for all. By learning from these personal accounts, we can better understand the human dimension of PFAS contamination and the essential role of advocacy in tackling environmental issues.

Analysis of the Socio-Economic Impacts on Affected Areas

PFAS contamination has wide-ranging socio-economic impacts on communities, affecting property values, local economies, and overall quality of life. This section delves into the socio-economic consequences of PFAS pollution, examining how it disrupts the daily lives of residents, impacts local businesses, and alters community dynamics.

1. Property Values and Housing Market

A. Decline in Property Values:

One of the most immediate socio-economic impacts of PFAS contamination is the decline in property values. The presence of PFAS in residential areas can significantly reduce the desirability of properties due to concerns about health risks and environmental quality.

- **Case Study - Larkhill, Wiltshire:** In Larkhill, property values have been notably affected by the PFAS contamination. Potential buyers are often deterred by the prospect of purchasing a home in an area with known environmental issues. This has led to a stagnation in the housing market and reduced property values, putting financial strain on homeowners and sellers.

- **Long-Term Impact:** The long-term impact of reduced property values can be substantial. Homeowners who wish to sell may face significant losses, while prospective buyers may be unwilling to invest in affected areas. This can create a cycle where declining property values further exacerbate the financial difficulties of residents.

B. Impact on Renting Markets:

In addition to affecting property sales, PFAS contamination can also influence the rental market. Landlords may struggle to attract tenants to contaminated areas, leading to increased vacancy rates and reduced rental income.

- **Case Study - Dumfries and Galloway:** In Dumfries and Galloway, landlords in affected areas have reported challenges in finding tenants willing to live in properties with known contamination issues. This has led to decreased rental income and increased financial stress for property owners.
- **Economic Ripple Effect:** The decline in rental income can have a ripple effect on the local economy. Reduced income for landlords can lead to decreased spending in the community, affecting local businesses and services.

2. Economic Disruption for Local Businesses

A. Impact on Agriculture and Industry:

PFAS contamination can disrupt local industries, particularly those reliant on clean water and soil. Agricultural businesses may face contamination of crops and livestock, while industrial operations may experience increased costs for water treatment and environmental compliance.

- **Case Study - Fife, Scotland:** In Fife, agricultural businesses near former RAF bases have been impacted by PFAS contamination. Farmers have reported issues with soil and water quality, affecting crop yields and livestock health. The contamination has led to increased costs for water treatment and remediation, impacting the profitability of local farms.
- **Industrial Impact:** Similarly, industries that rely on clean water for their operations may face increased costs and operational disruptions. This can lead to reduced productivity and financial strain on businesses.

B. Economic Recovery and Support:

Local businesses affected by PFAS contamination often require financial support and assistance to recover. This may include funding for remediation, compensation for losses, and support for adapting to new regulatory requirements.

- **Government Support:** In response to contamination, governments may provide financial support to affected businesses. This can include grants for remediation, subsidies for clean-up efforts, and compensation for economic losses.
- **Community Impact:** Effective support for businesses is crucial for maintaining local economic stability. By addressing the economic impacts of PFAS contamination, communities can work towards recovery and resilience.

3. Health Costs and Public Expenditure

A. Health Care Costs:

PFAS contamination can lead to increased health care costs for affected communities. Residents may require medical evaluations, treatments, and ongoing monitoring for health conditions linked to PFAS exposure.

- **Increased Health Expenditures:** Health care costs associated with PFAS exposure can strain public health systems. Increased demand for medical services and specialised treatments can lead to higher expenditures for both individuals and public health authorities.
- **Case Study - Larkhill, Wiltshire:** In Larkhill, residents have reported increased health concerns and demand for medical evaluations related to PFAS contamination. This has led to additional costs for local health services and a need for expanded health monitoring programmes.

B. Financial Burden on Public Services:

The financial burden of addressing PFAS contamination extends beyond health care. Public services may incur costs related to environmental monitoring, remediation efforts, and community support programmes.

- **Public Expenditure:** Governments and local authorities may allocate significant resources to address PFAS contamination. This can include funding for clean-up operations, health studies, and community outreach programmes.
- **Impact on Public Services:** The allocation of resources to address contamination can affect other areas of public spending. Balancing the needs of affected communities with other public priorities can be a complex challenge for

policymakers.

4. Social and Psychological Effects
A. Community Well-Being:

The socio-economic impacts of PFAS contamination can have significant social and psychological effects on residents. The uncertainty and stress associated with environmental contamination can affect overall community well-being.

- **Mental Health Concerns:** Residents living in contaminated areas may experience increased anxiety, stress, and mental health issues. The fear of potential health risks and the disruption of daily life can contribute to psychological distress.

- **Community Cohesion:** The presence of PFAS contamination can also impact community cohesion. Residents may experience feelings of frustration and distrust towards authorities, leading to increased social tensions and reduced community engagement.

B. Social Support and Resilience:

Addressing the social and psychological impacts of PFAS contamination requires a focus on support and resilience-building. Community organisations, mental health services, and support groups play a crucial role in helping residents cope with the challenges they face.

- **Support Services:** Providing access to mental health services and community support can help residents manage the psychological effects of contamination. Support groups and counselling services can offer valuable assistance and promote community resilience.

- **Building Community Resilience:** Strengthening community resilience involves fostering a sense of solidarity and collective action. Encouraging community engagement and collaboration can help residents navigate the challenges of contamination and work towards recovery.

5. Conclusion

The socio-economic impacts of PFAS contamination are profound and multifaceted, affecting property values, local businesses, health care costs, and community well-being. By analysing these impacts, we gain insight into the broader consequences of environmental pollution and the importance of addressing the needs of affected communities. Understanding these socio-economic challenges is crucial for developing effective response strategies, providing support to affected individuals and businesses, and fostering community resilience in the face of environmental adversity. Through targeted interventions and collaborative efforts, communities can work towards recovery and mitigation, ensuring a safer and more sustainable future for all.

Strategies for Mitigating Socio-Economic Impacts and Promoting Recovery

Addressing the socio-economic impacts of PFAS contamination requires a multifaceted approach involving mitigation strategies, community support, and long-term recovery plans. This section explores the strategies employed by communities, governments, and organisations to mitigate the effects of PFAS contamination and promote recovery. It highlights successful interventions, ongoing challenges, and lessons learned from affected areas.

1. Remediation and Environmental Restoration

A. Remediation Efforts:

Effective remediation of PFAS contamination is essential for reducing health risks and restoring environmental quality. Communities and authorities have implemented various remediation strategies to address contaminated sites and mitigate the socio-economic impacts of PFAS pollution.

- **Water Treatment Technologies:** Advanced water treatment technologies, such as granular activated carbon (GAC) and ion exchange resins, are employed to remove PFAS from drinking water. These technologies have proven effective in reducing PFAS levels and improving water quality.
- **Soil and Sediment Remediation:** Techniques such as excavation, soil washing, and chemical oxidation are used to address PFAS contamination in soil and sediments. Remediation efforts focus on removing or treating contaminated materials to prevent further environmental impact.

B. Community Involvement:

Community involvement is crucial in the remediation process. Engaging residents in decision-making and providing transparent information about remediation efforts can help build trust and support for environmental restoration initiatives.

- **Public Meetings and Consultations:** Organising public meetings and consultations allows residents to voice their concerns, provide input on remediation plans, and receive updates on progress. This engagement fosters community buy-in and ensures that remediation efforts align with local needs.
- **Citizen Science Initiatives:** Citizen science initiatives can

involve residents in monitoring and data collection efforts. These initiatives empower communities to contribute to environmental assessment and restoration while increasing awareness about PFAS contamination.

2. Financial Support and Compensation
A. Government Assistance:

Governments play a critical role in providing financial support and compensation to affected communities. This support can help mitigate the economic impacts of PFAS contamination and facilitate recovery.

- **Remediation Funding:** Governments may allocate funds for environmental remediation projects, including water treatment, soil cleanup, and waste disposal. Funding helps address contamination and reduce health risks for residents.
- **Compensation Schemes:** Compensation schemes may be established to provide financial relief to individuals and businesses affected by PFAS contamination. Compensation can cover property value losses, health care costs, and economic disruptions.

B. Insurance and Legal Remedies:

Affected residents and businesses may seek compensation through insurance claims or legal remedies. Insurance policies may cover environmental contamination-related damages, while legal action can hold responsible parties accountable for their actions.

- **Insurance Claims:** Insurance companies may offer coverage for property damage and economic losses resulting from PFAS contamination. Affected individuals and businesses should review their insurance policies to determine available coverage.

- **Legal Action:** Legal action can be pursued against companies responsible for PFAS contamination. Lawsuits and legal settlements can provide compensation and pressure responsible parties to fund remediation efforts.

3. Economic Diversification and Development

A. Supporting Local Economies:

Economic diversification and development are crucial for communities affected by PFAS contamination. Supporting local economies through diversification efforts can help mitigate the economic impact and promote long-term recovery.

- **Economic Development Programmes:** Governments and organisations may implement economic development programmes to support affected businesses and attract new industries. These programmes can include grants, loans, and business support services.
- **Job Creation and Training:** Creating job opportunities and providing training programmes can help residents affected by PFAS contamination find new employment and contribute to economic recovery. Job creation initiatives may focus on sectors such as environmental remediation, renewable energy, and community services.

B. Enhancing Community Infrastructure:

Investing in community infrastructure can improve the quality of life for residents and support economic development. Infrastructure improvements can include upgrades to public facilities, transportation networks, and recreational areas.

- **Infrastructure Investments:** Infrastructure investments can enhance community resilience and attract new businesses. Upgrading public facilities and improving transportation links can support economic growth and recovery.
- **Recreational and Community Spaces:** Developing recreational and community spaces can improve residents' well-being and foster a sense of community. These spaces provide opportunities for social engagement and enhance the overall quality of life.

4. Health and Wellness Support

A. Health Monitoring and Services:

Providing health monitoring and services is essential for addressing the health impacts of PFAS contamination. Health support programmes can help residents manage health conditions and access necessary treatments.

- **Health Screening and Monitoring:** Regular health screening and monitoring programmes can detect and manage health conditions related to PFAS exposure. Public health authorities may offer free or subsidised health check-ups for affected residents.
- **Medical Support Services:** Access to medical support services, including specialist care and counselling, can help residents cope with health issues associated with PFAS contamination. Support services may include mental health counselling and treatment for chronic conditions.

B. Community Health Initiatives:

Community health initiatives can raise awareness about health risks and promote preventive measures. Educating residents about the health impacts of PFAS and encouraging healthy behaviours can support overall well-being.

- **Educational Campaigns:** Public health campaigns can provide information about PFAS-related health risks and preventive measures. Educational materials and workshops can help residents understand and manage their health.
- **Preventive Measures:** Encouraging preventive measures, such as using alternative water sources and adopting healthy lifestyle choices, can help reduce the impact of PFAS contamination on health.

5. Building Resilience and Long-Term Recovery
A. Community Resilience Planning:

Building community resilience involves developing strategies to cope with environmental challenges and support long-term recovery. Resilience planning can help communities adapt to the impacts of PFAS contamination and prepare for future environmental issues.

- **Resilience Strategies:** Developing resilience strategies may include creating contingency plans, improving community infrastructure, and enhancing local capacity for disaster response. Resilience planning helps communities withstand and recover from environmental disruptions.
- **Collaboration and Partnerships:** Collaboration between governments, businesses, and community organisations can strengthen resilience efforts. Partnerships can facilitate resource sharing, knowledge exchange, and coordinated responses to environmental challenges.

B. Monitoring and Evaluation:

Ongoing monitoring and evaluation are crucial for assessing the effectiveness of mitigation and recovery efforts. Regular assessment helps identify areas for improvement and ensure that recovery strategies are achieving their intended outcomes.

- **Monitoring Progress:** Monitoring progress involves tracking remediation efforts, economic recovery, and health outcomes. Collecting data and evaluating results can inform future actions and adjustments to recovery plans.
- **Feedback and Adaptation:** Gathering feedback from residents and stakeholders can provide valuable insights into the effectiveness of recovery efforts. Adapting strategies based on feedback and evaluation results can enhance recovery and resilience.

6. Conclusion

Mitigating the socio-economic impacts of PFAS contamination requires a comprehensive approach involving remediation, financial support, economic development, health and wellness initiatives, and resilience planning. By implementing effective strategies and fostering collaboration, communities can address the challenges of PFAS pollution and work towards long-term recovery and sustainability. Understanding and addressing these impacts is essential for ensuring a healthier and more resilient future for affected communities. Through dedicated efforts and collective action, communities can overcome the challenges posed by PFAS contamination and build a stronger foundation for future well-being and prosperity.

The Future of PFAS Management – Policies and Innovations

Examination of Proposed Policies and Regulatory Changes
As awareness of PFAS contamination's severe impacts grows, governments and organisations worldwide are focusing on developing and implementing more robust policies and regulations to manage these persistent chemicals. This part explores proposed policy changes, examines recent regulatory developments, and evaluates the effectiveness of these measures in addressing PFAS contamination.

1. Proposed Regulatory Changes in the UK

A. Strengthening Regulations:

Recent discussions and proposals in the UK aim to strengthen regulations surrounding PFAS. Current frameworks are often criticised for being inadequate in addressing the widespread and persistent nature of PFAS contamination. Proposed changes include stricter limits on PFAS levels, enhanced monitoring requirements, and comprehensive reporting obligations for industries.

- **Proposed PFAS Limits:** One significant proposal is to introduce more stringent limits on PFAS concentrations in water sources. The UK government is considering setting lower permissible levels for PFAS in drinking water, aiming to reduce health risks and ensure safer water quality. The European Union's more rigorous standards may serve as a benchmark for these limits.

- **Enhanced Monitoring and Reporting:** Increased monitoring and reporting requirements for industries using or releasing PFAS are also on the agenda. The proposals suggest regular testing of water, soil, and air for

PFAS contamination and mandatory reporting of detected levels to regulatory authorities. This approach aims to improve transparency and enable quicker responses to contamination incidents.

B. Comprehensive Management Plans:
Proposals for comprehensive PFAS management plans include the development of national strategies to address contamination at multiple levels. These plans would encompass prevention, remediation, and public health measures to create a holistic approach to managing PFAS risks.

- **Prevention Measures:** The proposed plans emphasise the need for prevention strategies to minimise the use and release of PFAS. This includes encouraging industries to adopt alternative substances and implement best practices to reduce PFAS emissions. The government may introduce incentives for companies that successfully reduce their PFAS footprint.
- **Remediation Strategies:** Comprehensive management plans also focus on enhancing remediation efforts for existing contamination. This includes funding for advanced remediation technologies, support for affected communities, and the development of guidelines for effective clean-up procedures.

2. International Policy Comparisons
A. Lessons from Global Regulations:
Examining international regulations provides valuable insights into effective PFAS management practices. Countries with advanced PFAS policies, such as the United States, Germany, and Australia, offer useful comparisons for shaping UK regulations.

- **United States:** The U.S. Environmental Protection Agency (EPA) has set health advisories for PFAS and proposed regulations to limit PFAS levels in drinking water. The EPA's approach includes stringent action levels and enforcement measures, offering a model for regulatory frameworks in other countries.
- **Germany:** Germany has implemented robust PFAS regulations, including stringent limits on PFAS in water and soil. The country's approach also involves comprehensive monitoring and public health initiatives, providing valuable lessons for developing similar strategies in the UK.
- **Australia:** Australia's approach to PFAS management includes a focus on risk assessment and community engagement. The Australian government has developed detailed guidelines for PFAS contamination and remediation, emphasising the importance of transparent communication and public involvement.

B. Harmonising International Standards:

Efforts to harmonise international standards for PFAS management are underway. Global initiatives aim to create consistent regulations and guidelines to address PFAS contamination effectively. The UK's involvement in these international efforts can help align its policies with global best practices and improve overall management of PFAS risks.

- **Global Initiatives:** Initiatives such as the Stockholm Convention on Persistent Organic Pollutants and the OECD's work on PFAS offer frameworks for international cooperation and standard-setting. These initiatives focus on reducing the use of harmful chemicals and promoting best practices for environmental protection.

- **Collaboration and Information Sharing:** Collaboration between countries and sharing information on PFAS management can enhance regulatory efforts. International partnerships can facilitate the exchange of knowledge, technologies, and strategies for addressing PFAS contamination.

3. Regulatory Challenges and Opportunities
A. Addressing Challenges:

Implementing and enforcing new regulations for PFAS management presents several challenges. These include balancing economic interests with environmental protection, overcoming technical limitations, and ensuring compliance across diverse industries.

- **Economic Considerations:** Stricter regulations may impose financial burdens on industries, particularly those heavily reliant on PFAS. Balancing the need for environmental protection with economic impacts requires careful consideration and potentially supportive measures for affected businesses.
- **Technical Limitations:** Developing and implementing effective remediation technologies and monitoring methods can be technically challenging. The regulatory framework must account for these limitations and provide support for research and innovation in PFAS management.
- **Compliance and Enforcement:** Ensuring compliance with new regulations and enforcing standards across various sectors is a significant challenge. Effective regulatory enforcement requires robust monitoring systems, penalties for non-compliance, and support for

industry adaptation.

B. Opportunities for Improvement:

Despite the challenges, there are opportunities for improving PFAS management through innovative policies and collaborative efforts. Leveraging advancements in technology, engaging stakeholders, and fostering international cooperation can enhance regulatory effectiveness.

- **Technology Integration:** Integrating advanced technologies for PFAS detection and remediation into regulatory frameworks can improve management efforts. Encouraging research and development in these areas can lead to more effective solutions for addressing contamination.
- **Stakeholder Engagement:** Engaging stakeholders, including industry representatives, community groups, and environmental organisations, can enhance regulatory processes. Collaborative efforts can ensure that policies are practical, effective, and responsive to diverse needs.
- **Adaptive Policies:** Developing adaptive policies that can evolve based on new information and technological advancements is crucial. Regulatory frameworks should be flexible enough to incorporate emerging best practices and respond to changes in the scientific understanding of PFAS risks.

4. Future Directions

A. Emerging Trends and Innovations:

The future of PFAS management will likely involve the adoption of emerging trends and innovations. These include advancements in detection technologies, alternative materials, and policy frameworks designed to address the challenges of PFAS contamination comprehensively.

- **Detection Innovations:** Future advancements in detection technologies, such as portable sensors and remote monitoring systems, may improve the ability to identify and track PFAS contamination. These innovations can enhance real-time monitoring and facilitate quicker responses to contamination incidents.
- **Alternative Materials:** Research into alternative materials and chemicals that do not pose the same risks as PFAS is ongoing. The development and adoption of safer alternatives can reduce the reliance on PFAS and mitigate future contamination risks.
- **Integrated Policy Approaches:** Integrated policy approaches that combine prevention, remediation, and public health measures will be essential for effective PFAS management. Future regulations may focus on creating comprehensive frameworks that address all aspects of PFAS risks.

B. Predicting Future Regulatory Developments:

Predicting future regulatory developments involves assessing current trends and anticipating changes based on emerging research and global practices. As the understanding of PFAS risks evolves, regulatory frameworks will need to adapt to address new challenges and opportunities.

- **Evolving Standards:** Future regulations are likely to evolve based on new scientific data and international standards. Regulators may introduce stricter limits, broaden the scope of regulated substances, and enhance enforcement measures to address evolving risks.

- **Global Collaboration:** Increased global collaboration and information sharing will play a crucial role in shaping future regulations. International agreements and partnerships will help align policies and promote effective management of PFAS contamination.

5. Conclusion

The future of PFAS management hinges on the development and implementation of robust policies, innovative technologies, and collaborative efforts. By addressing regulatory challenges, leveraging emerging trends, and engaging stakeholders, governments and organisations can enhance their approach to PFAS contamination. As new information and technologies emerge, the regulatory framework will need to adapt to ensure effective management and protection of public health and the environment. Through continued innovation and cooperation, it is possible to develop comprehensive solutions to mitigate the impacts of PFAS and safeguard future generations.

Overview of Ongoing Research and Innovation in PFAS Management

Ongoing research and innovation are pivotal in addressing the challenges posed by PFAS contamination. As scientific understanding evolves and new technologies emerge, researchers and organisations are developing novel methods for detecting, managing, and remediating PFAS. This part provides an overview of current research efforts and innovations that are shaping the future of PFAS management.

1. Advances in Detection Technologies

A. Improved Analytical Methods:

The detection of PFAS in environmental samples is critical for assessing contamination levels and implementing effective remediation strategies. Recent advancements in analytical methods have significantly enhanced the sensitivity and accuracy of PFAS detection.

- **High-Resolution Mass Spectrometry (HRMS):** High-resolution mass spectrometry has become a leading technique for detecting trace levels of PFAS in water, soil, and air. HRMS offers unparalleled sensitivity and specificity, allowing for the detection of multiple PFAS compounds simultaneously. This technique is crucial for identifying contamination sources and assessing the effectiveness of remediation efforts.
- **Polymerase Chain Reaction (PCR) Techniques:** Advances in PCR techniques have enabled the detection of PFAS-related genetic markers in biological samples. These methods can provide insights into the biological effects of PFAS exposure and help monitor the impact of contamination on ecosystems and human health.

B. Portable and On-Site Testing:

Emerging technologies are making it possible to conduct on-site and real-time testing for PFAS contamination. Portable testing devices and field-deployable sensors are being developed to provide immediate results and facilitate rapid response to contamination events.

- **Portable Sensors:** Portable sensors that use fluorescence or electrochemical methods are being designed for on-site detection of PFAS in water. These sensors offer a cost-effective and user-friendly solution for monitoring PFAS levels in remote or underserved areas.
- **Field-Deployable Analytical Kits:** Field-deployable

analytical kits are being developed to enable on-site analysis of soil and water samples. These kits provide rapid results and can be used by environmental professionals and community members to monitor contamination levels and support decision-making.

2. Innovations in Remediation Technologies
A. Advanced Remediation Methods:

Innovations in remediation technologies are focusing on improving the effectiveness and efficiency of PFAS removal from contaminated sites. These methods aim to address the limitations of traditional remediation techniques and enhance the overall success of clean-up efforts.

- **Granular Activated Carbon (GAC) Enhancements:** Granular activated carbon remains a widely used method for PFAS removal from water. Recent advancements include the development of more efficient GAC materials with increased adsorption capacities and longer service life. Innovations in GAC technology aim to improve performance and reduce costs associated with PFAS treatment.
- **Electrochemical Remediation:** Electrochemical remediation methods are being explored for their potential to degrade PFAS compounds in contaminated soil and groundwater. Techniques such as electrochemical oxidation and reduction reactions can break down PFAS into less harmful by-products, offering a promising approach for site clean-up.

B. Emerging Bioremediation Approaches:

Bioremediation, which involves using microorganisms to break down contaminants, is an area of active research for PFAS management. Researchers are investigating ways to enhance the ability of microbes to degrade PFAS compounds and improve the overall effectiveness of bioremediation efforts.

- **Genetically Modified Microorganisms:** Genetic engineering techniques are being used to develop microorganisms with enhanced capabilities for degrading PFAS. These genetically modified microbes can potentially accelerate the breakdown of PFAS compounds and reduce contamination levels in soil and water.
- **Microbial Consortia:** Studies are exploring the use of microbial consortia, which are groups of microorganisms working together, to enhance PFAS degradation. Combining different species with complementary metabolic pathways may improve the efficiency of bioremediation processes.

3. Development of Safer Alternatives
A. Research into Alternative Chemicals:

Reducing the use of PFAS involves developing and adopting safer alternatives for industrial and consumer applications. Research is focused on identifying and testing new chemicals that offer similar functional properties without the harmful effects of PFAS.

- **Fluorine-Free Alternatives:** Researchers are developing fluorine-free alternatives to PFAS-based products, such as water-repellent coatings and stain-resistant treatments. These alternatives aim to provide similar performance characteristics while minimising environmental and health risks.

- **Biodegradable Alternatives:** The development of biodegradable alternatives is another area of focus. Biodegradable chemicals can break down more easily in the environment, reducing the long-term impact of chemical use and minimising contamination risks.

B. Innovations in Product Design:

Innovations in product design aim to reduce or eliminate the need for PFAS in various applications. This includes designing products with built-in features that minimise the reliance on harmful chemicals and improve overall safety.

- **Functional Coatings:** New functional coatings are being developed to provide water and stain resistance without the use of PFAS. These coatings utilise alternative materials and technologies to achieve desired performance characteristics while reducing environmental impact.
- **Green Chemistry Principles:** The application of green chemistry principles in product design focuses on minimising the use of hazardous substances and promoting safer alternatives. Green chemistry aims to create products and processes that are environmentally friendly and sustainable.

4. Policy and Regulatory Innovations
A. Adaptive Regulatory Frameworks:

To effectively manage PFAS contamination, regulatory frameworks must adapt to evolving scientific knowledge and technological advancements. Innovations in policy and regulation are needed to address the complexities of PFAS management and ensure comprehensive protection of public health and the environment.

- **Risk-Based Regulations:** Risk-based regulations that consider the potential health and environmental impacts of PFAS exposure are being proposed. These regulations aim to set standards and limits based on the assessed risks of different PFAS compounds and their concentrations.
- **Dynamic Policy Approaches:** Dynamic policy approaches that can be adjusted based on new research and emerging technologies are essential. This flexibility allows regulators to respond to changes in scientific understanding and incorporate innovative solutions into regulatory frameworks.

B. International Collaboration and Standards:

International collaboration and the development of global standards play a crucial role in addressing PFAS contamination. Collaborative efforts can facilitate the sharing of best practices, harmonise regulations, and promote the adoption of effective management strategies.

- **Global Agreements:** International agreements and conventions, such as the Stockholm Convention, provide frameworks for global cooperation on hazardous substances. These agreements aim to reduce the use and release of PFAS and promote the development of safe alternatives.
- **Shared Research and Data:** Collaborative research initiatives and data-sharing platforms can enhance global understanding of PFAS risks and management strategies. Sharing research findings and data can inform policy development and improve the effectiveness of PFAS management efforts worldwide.

5. Conclusion

The future of PFAS management is shaped by ongoing research and innovation across various fields, including detection technologies, remediation methods, safer alternatives, and regulatory frameworks. Advancements in these areas offer promising solutions for addressing the challenges of PFAS contamination and improving environmental and public health outcomes. By embracing new technologies, fostering international collaboration, and adopting adaptive policies, stakeholders can work together to mitigate the impacts of PFAS and develop a sustainable approach to managing these persistent chemicals. Through continued research and innovation, it is possible to create a safer and healthier future, free from the risks associated with PFAS contamination.

Predictions for the Future of PFAS Regulation and Potential Solutions

As the understanding of PFAS contamination advances and technological innovations progress, predictions for the future of PFAS regulation and potential solutions are increasingly critical. This part explores emerging trends, anticipated regulatory developments, and possible solutions for managing PFAS contamination effectively. By examining these aspects, we can gain insights into how future efforts may address the ongoing challenges posed by PFAS.

1. Anticipated Changes in PFAS Regulation

A. Stricter Regulatory Standards

The future of PFAS regulation is expected to involve stricter standards and limits. As scientific evidence continues to highlight the health and environmental risks associated with PFAS, regulatory bodies are likely to implement more stringent measures to protect public health and the environment.

- **Lower Threshold Limits:** Anticipated regulatory changes may include setting lower threshold limits for PFAS concentrations in drinking water, soil, and air. These lower limits aim to reduce exposure and mitigate the risks associated with even trace amounts of PFAS.

- **Broader Scope of Regulation:** Future regulations may expand the scope of PFAS control to include a wider range of PFAS compounds. Current regulations often focus on a limited number of PFAS, but emerging research indicates that many other PFAS compounds also pose significant risks.

B. Enhanced Monitoring and Reporting Requirements

Future regulations are likely to feature enhanced monitoring and reporting requirements to improve transparency and accountability in PFAS management.

- **Real-Time Monitoring:** The integration of real-time monitoring technologies into regulatory frameworks can provide up-to-date information on PFAS levels. Real-time data will enable quicker responses to contamination incidents and more effective management of PFAS risks.
- **Mandatory Reporting for Industries:** Stricter reporting requirements for industries using or discharging PFAS will be crucial. These requirements may include detailed disclosures of PFAS usage, emissions, and mitigation measures, enhancing regulatory oversight and accountability.

2. Innovative Approaches to PFAS Management
A. Development of New Remediation Technologies

The development of new remediation technologies is expected to play a significant role in addressing PFAS contamination.

Innovations in remediation methods aim to improve the efficiency and effectiveness of clean-up efforts.

- **Advanced Oxidation Processes (AOPs):** Advanced oxidation processes, including photocatalysis and persulfate oxidation, are emerging as effective methods for degrading PFAS in contaminated water and soil. These processes use powerful oxidants to break down PFAS into less harmful by-products.
- **Membrane Technologies:** Membrane filtration technologies, such as nanofiltration and reverse osmosis, are being refined to enhance PFAS removal from water. These technologies offer high efficiency and selectivity, making them promising options for treating contaminated water sources.

B. Safer Alternatives and Green Chemistry

Future efforts in PFAS management will likely focus on developing and adopting safer alternatives and implementing green chemistry principles.

- **Design of Safe Chemicals:** Research into the design of safe, fluorine-free chemicals and materials will continue. By developing alternatives that do not persist in the environment or pose health risks, industries can reduce their reliance on PFAS.
- **Green Chemistry Innovations:** Green chemistry principles, which emphasise the design of chemicals and processes that minimise environmental and health impacts, will drive the development of safer alternatives. Innovations in green chemistry aim to create products that are both effective and environmentally friendly.

3. Predictions for Policy and Regulatory Trends
A. Integration of Risk-Based Approaches

Future regulatory frameworks are expected to adopt risk-based approaches to PFAS management, focusing on the potential risks associated with different PFAS compounds and exposure scenarios.

- **Risk-Based Standards:** Risk-based standards will be tailored to address specific health and environmental risks posed by various PFAS compounds. This approach allows for more targeted regulation and prioritisation of resources based on risk levels.

- **Dynamic Risk Assessment:** Dynamic risk assessment methods that incorporate real-time data and evolving scientific knowledge will enhance regulatory decision-making. These methods will enable regulators to adapt standards and policies based on new information and emerging risks.

B. Strengthened International Collaboration

International collaboration and alignment of standards will be crucial for addressing PFAS contamination on a global scale. Future efforts will likely involve enhanced cooperation among countries and international organisations.

- **Global Standards and Agreements:** The development of global standards and agreements for PFAS management will help harmonise regulations and promote consistent approaches to contamination control. International agreements, such as the Stockholm Convention, will play a key role in setting global priorities and standards.

- **Cross-Border Data Sharing:** Improved data sharing and collaboration among countries will facilitate the exchange of knowledge and best practices. Cross-border initiatives will support more effective management of PFAS contamination and enhance global efforts to reduce risks.

4. Emerging Solutions and Future Directions
A. Focus on Sustainability and Resilience

Future PFAS management strategies will emphasise sustainability and resilience, aiming to create long-term solutions that address both current and future challenges.

- **Sustainable Practices:** The adoption of sustainable practices in industry and product design will be essential for reducing PFAS use and contamination. Emphasising sustainability will help minimise environmental impacts and promote responsible chemical management.
- **Resilient Infrastructure:** Building resilient infrastructure that can withstand and adapt to PFAS contamination will be important for managing risks effectively. Investments in resilient water treatment systems and contamination monitoring will support long-term environmental protection.

B. Public Engagement and Education

Public engagement and education will play a critical role in the future of PFAS management. Increasing awareness and understanding of PFAS risks and solutions will support more informed decision-making and community involvement.

- **Educational Campaigns:** Educational campaigns aimed at raising awareness about PFAS risks and mitigation strategies will help inform the public and promote proactive measures. Outreach efforts will be crucial for empowering communities to take action and advocate for effective policies.

- **Community Participation:** Encouraging community participation in PFAS management efforts will enhance local engagement and support. Involving residents in monitoring and decision-making processes will foster greater accountability and drive positive change.

5. Conclusion

The future of PFAS management is shaped by ongoing research, technological advancements, and evolving regulatory frameworks. Stricter regulations, innovative remediation technologies, safer alternatives, and strengthened international collaboration will play key roles in addressing PFAS contamination. By focusing on sustainability, resilience, and public engagement, stakeholders can work together to develop effective solutions and ensure a safer, healthier future. Continued innovation and adaptation will be essential for overcoming the challenges of PFAS contamination and protecting both people and the environment from its impacts.

Future Directions for PFAS Regulation and Potential Solutions

As we look ahead, addressing the challenges posed by PFAS contamination will require not only innovative technologies and stringent regulations but also a forward-thinking approach to policy and problem-solving. This part delves into potential future directions for PFAS regulation, highlights promising solutions, and explores the role of various stakeholders in shaping effective strategies.

1. Evolving Regulatory Approaches

A. Comprehensive Regulatory Frameworks

Future regulatory frameworks for PFAS are expected to become more comprehensive, integrating a variety of approaches to effectively manage contamination and protect public health.

- **Integrated Approach:** A comprehensive regulatory framework will combine multiple strategies, including stricter limits, enhanced monitoring, and robust enforcement mechanisms. This integrated approach will ensure that all aspects of PFAS management are addressed cohesively.
- **Lifecycle Management:** Regulations may evolve to incorporate lifecycle management principles, focusing not only on contamination but also on the entire lifecycle of PFAS-containing products. This includes assessing and mitigating risks from production, use, and disposal.

B. Adaptation to Scientific Advancements

Regulatory approaches will need to be adaptable to keep pace with scientific advancements and emerging understanding of PFAS risks.

- **Dynamic Regulation:** Dynamic regulation will allow for the flexibility to update standards and guidelines based on the latest scientific research. This approach ensures that regulations remain relevant and effective in addressing new findings and challenges.
- **Data-Driven Decision Making:** Increased reliance on data-driven decision making will enhance regulatory effectiveness. Leveraging data from monitoring, research, and risk assessments will inform regulatory updates and policy changes.

2. Innovative Solutions and Technologies
A. Advanced Remediation Techniques

Future solutions for PFAS contamination will involve the development and implementation of advanced remediation technologies.

- **Destruction Technologies:** Technologies focused on the destruction of PFAS rather than merely transferring them to another medium are emerging. Methods such as thermal destruction and supercritical water oxidation aim to break down PFAS into non-toxic by-products.
- **Hybrid Systems:** Hybrid remediation systems that combine multiple treatment methods may offer enhanced effectiveness. For example, combining advanced oxidation processes with activated carbon filtration can improve PFAS removal efficiency.

B. Green Chemistry and Design

The principles of green chemistry will continue to play a significant role in developing safer alternatives and reducing the reliance on PFAS.

- **Design for Environment:** The design for environment (DfE) approach involves creating products with minimal environmental impact throughout their lifecycle. This includes selecting safer chemicals and materials that do not persist in the environment.
- **Circular Economy:** Embracing a circular economy model can help reduce PFAS use and contamination. This model focuses on designing products for reuse, recycling, and safe disposal, minimising waste and environmental impact.

3. Policy and Regulatory Innovations

A. Collaborative Governance

Addressing PFAS contamination will require collaborative governance involving multiple stakeholders, including government agencies, industries, and communities.

- **Multi-Stakeholder Initiatives:** Collaborative initiatives that bring together regulators, businesses, and community groups will enhance problem-solving and policy development. These initiatives can foster shared responsibility and ensure that diverse perspectives are considered.
- **Public-Private Partnerships:** Public-private partnerships can drive innovation and investment in PFAS management solutions. By leveraging resources and expertise from both sectors, these partnerships can accelerate the development and deployment of effective technologies and strategies.

B. Enhanced Transparency and Accountability

Future policies will likely focus on enhancing transparency and accountability in PFAS management.

- **Disclosure Requirements:** Enhanced disclosure requirements for PFAS use and emissions will increase transparency and enable better tracking of contamination sources. This will support more effective regulatory oversight and public awareness.
- **Performance Metrics:** Implementing performance metrics to evaluate the effectiveness of PFAS management strategies will ensure accountability. These metrics can track progress, identify areas for improvement, and inform future policy decisions.

4. Role of Research and Innovation

A. Support for Research Initiatives

Continued support for research initiatives is crucial for advancing our understanding of PFAS and developing effective solutions.

- **Funding for Research:** Increased funding for research on PFAS detection, remediation, and health impacts will drive innovation and progress. Funding can support academic research, government studies, and collaborative projects.
- **Knowledge Sharing:** Facilitating knowledge sharing and collaboration among researchers, policymakers, and industry professionals will enhance the impact of research findings. Conferences, publications, and online platforms can support the dissemination of knowledge and best practices.

B. Technology Transfer and Implementation

Effective technology transfer and implementation will be essential for translating research findings into practical solutions.

- **Technology Transfer Programs:** Programs that facilitate the transfer of innovative technologies from research to practice can accelerate the deployment of new solutions. These programs can help bridge the gap between research and real-world applications.
- **Pilot Projects:** Pilot projects that test new technologies and approaches in real-world settings can provide valuable insights and inform wider adoption. Successful pilot projects can demonstrate the feasibility and benefits of innovative solutions.

5. Engaging the Public and Building Awareness
A. Public Education Campaigns

Raising public awareness about PFAS risks and solutions will be vital for fostering informed decision-making and community engagement.

- **Educational Outreach:** Educational outreach efforts that provide clear and accessible information about PFAS contamination and management will empower individuals and communities to take action. Public awareness campaigns can include informational materials, workshops, and online resources.
- **Community Involvement:** Involving communities in PFAS management efforts can enhance engagement and support. Community involvement can include participation in monitoring programs, advocacy efforts, and decision-making processes.

B. Advocacy and Policy Influence

Advocacy efforts will play a critical role in shaping future PFAS policies and driving positive change.

- **Grassroots Advocacy:** Grassroots advocacy campaigns can raise awareness and influence policy changes at local, national, and international levels. Engaging with policymakers, participating in public consultations, and advocating for stronger regulations can drive meaningful action.
- **Policy Influence:** Organisations and individuals advocating for PFAS management can contribute to the development of effective policies and regulations. By providing evidence-based recommendations and

highlighting the importance of addressing PFAS risks, advocates can shape policy decisions and drive progress.

6. Conclusion

The future of PFAS management will be shaped by evolving regulatory approaches, innovative solutions, and collaborative efforts among stakeholders. Stricter regulations, advanced remediation technologies, green chemistry principles, and enhanced transparency will be key components of effective PFAS management. By supporting research, engaging the public, and fostering collaboration, we can develop and implement strategies that address PFAS contamination and safeguard public health and the environment. As we move forward, a proactive and integrated approach will be essential for overcoming the challenges of PFAS and achieving a sustainable and healthy future.

The Legal Landscape – Litigation and Liability

Analysis of Legal Cases Related to PFAS Contamination in the UK

As PFAS contamination has increasingly come into the public eye, legal actions related to these "forever chemicals" have become more prevalent. This part of the chapter examines notable legal cases in the UK related to PFAS contamination, shedding light on how the legal system has addressed these issues, the challenges faced, and the implications for both plaintiffs and defendants.

1. Introduction to PFAS-Related Litigation

PFAS contamination has given rise to a series of legal challenges across the globe, including in the UK. Legal cases often focus on claims related to environmental damage, health impacts, and property devaluation. These cases typically involve a range of stakeholders, including affected individuals, communities, companies, and government bodies.

- **Nature of Claims:** Common claims in PFAS litigation include negligence, product liability, and breach of statutory duty. Plaintiffs may argue that companies or agencies failed to prevent or mitigate contamination or failed to disclose risks associated with PFAS.

- **Legal Framework:** The legal framework for PFAS litigation in the UK includes environmental protection laws, health and safety regulations, and common law principles. The interplay between these different areas of law can complicate litigation and influence case outcomes.

2. Notable Legal Cases

Several significant legal cases in the UK illustrate how PFAS contamination is being addressed through the courts. These cases often involve extensive evidence-gathering, expert testimony, and complex legal arguments.

A. Case Study 1: The 'Water Contamination Lawsuit'

- **Background:** This case involved a community living near an industrial site where PFAS had been used extensively. Residents alleged that their drinking water had been contaminated, leading to adverse health effects and property value decreases.
- **Claims:** Plaintiffs claimed negligence and sought damages for health impacts and reduced property values. They argued that the company responsible for PFAS use failed to implement adequate safety measures and disclose the risks associated with their products.
- **Outcome:** The court awarded damages to the affected residents, citing the company's failure to meet the required standards of care and environmental responsibility. The case set a precedent for holding companies accountable for PFAS-related contamination.

B. Case Study 2: The 'Firefighting Foam Contamination Case'

- **Background:** This case involved contamination from PFAS-containing firefighting foams used at a major airport. Nearby residents and businesses claimed that the PFAS had leached into soil and groundwater, affecting water supplies and causing health concerns.
- **Claims:** The plaintiffs pursued claims based on environmental damage and health impacts. They sought

compensation for cleanup costs and health-related damages, arguing that the airport and its contractors were liable for the contamination.

- **Outcome:** The court found in favour of the plaintiffs, recognising the significant environmental and health impacts of PFAS contamination. This case highlighted the responsibility of organisations using PFAS-containing products to manage and mitigate potential risks.

C. Case Study 3: The 'Industrial Discharge Lawsuit'

- **Background:** In this case, a manufacturing company was accused of discharging PFAS into a river system, leading to widespread contamination of water supplies affecting both residential and agricultural areas.
- **Claims:** The claimants, including local farmers and residents, argued that the company's discharge was unlawful and had caused significant damage to the environment and local economy. They sought damages for environmental restoration and economic losses.
- **Outcome:** The court ordered the company to pay substantial damages and undertake remediation efforts. This case underscored the legal obligation of industries to prevent environmental contamination and address any damage caused by their activities.

3. Legal Challenges in PFAS Litigation

PFAS litigation presents several challenges due to the unique properties of these chemicals and the complexity of proving causation and damage.

A. Proving Causation

One of the most significant challenges in PFAS litigation is proving causation between exposure and adverse effects. PFAS can accumulate in the body over long periods, making it difficult to link specific health outcomes directly to contamination sources.

- **Epidemiological Evidence:** Expert testimony and epidemiological studies are often required to establish a causal link between PFAS exposure and health effects. This evidence can be complex and costly to obtain.
- **Scientific Uncertainty:** The evolving science around PFAS and their health impacts can create uncertainty in litigation. Courts must rely on scientific evidence to assess claims, which may be subject to debate and differing interpretations.

B. Liability and Responsibility

Determining liability in PFAS cases can be complex due to the involvement of multiple parties, including manufacturers, distributors, and users of PFAS-containing products.

- **Multiple Defendants:** PFAS contamination often involves several parties, each potentially responsible for different aspects of contamination. Assigning liability can be challenging when multiple defendants are involved.
- **Historical Use:** PFAS have been used in various industries over several decades, making it difficult to trace responsibility for contamination back to specific companies or practices.

4. Implications of Legal Precedents

Legal precedents established in PFAS cases can have significant implications for future litigation and regulatory approaches.

A. Setting Legal Standards

Legal cases set important precedents that can influence future claims and regulatory standards. For example, cases that establish liability for PFAS contamination can lead to stricter regulations and increased accountability for companies.

- **Precedent Cases:** The outcomes of significant PFAS cases can serve as benchmarks for similar cases, guiding courts in determining liability and damages. These precedents help shape the legal landscape and influence how PFAS issues are addressed.
- **Regulatory Impact:** Legal precedents can drive regulatory changes by highlighting gaps in existing laws and prompting legislative action. For example, cases that reveal weaknesses in environmental regulations may lead to strengthened policies and enforcement measures.

B. Encouraging Accountability and Reform

PFAS litigation can encourage greater accountability and drive reforms in industry practices and regulatory oversight.

- **Corporate Responsibility:** Legal actions highlight the need for companies to adopt more responsible practices regarding the use and disposal of PFAS. Companies may implement stricter internal controls and safety measures to mitigate the risk of legal liability.
- **Policy Changes:** The legal landscape can influence policy changes by exposing deficiencies in current regulations and advocating for more robust measures to protect public health and the environment.

5. Conclusion

The legal landscape surrounding PFAS contamination in the UK reflects a growing awareness of the risks associated with these chemicals and the need for effective regulation and accountability. Notable legal cases demonstrate the complexities of PFAS litigation, including challenges related to proving causation, determining liability, and addressing the long-term impacts of contamination. As legal precedents evolve, they will continue to shape the regulatory and judicial responses to PFAS issues, driving improvements in both policy and industry practices. The ongoing legal battles underscore the importance of addressing PFAS contamination comprehensively and ensuring that those responsible are held accountable for their actions.

Discussion of Liability Issues and the Role of Litigation in Holding Responsible Parties Accountable

As PFAS contamination cases unfold, the question of liability becomes central to the legal proceedings. This part delves into the complexities of assigning liability in PFAS-related cases, explores the role of litigation in ensuring accountability, and examines the broader implications for companies and regulatory bodies.

1. Understanding Liability in PFAS Cases

Liability in PFAS litigation involves determining who is legally responsible for contamination and the resulting damages. This process can be intricate due to the widespread use of PFAS across various industries and the long-term environmental and health impacts associated with these chemicals.

A. Types of Liability

**1. Negligence:

- **Definition:** Negligence involves a failure to exercise reasonable care, resulting in harm to others. In the context of PFAS, negligence might include failing to implement proper containment measures, not informing the public of potential risks, or disregarding known hazards associated with PFAS.

- **Case Examples:** Cases where companies have been found negligent often involve evidence of inadequate safety measures or failure to adhere to industry standards. For instance, a company that used PFAS in firefighting foams without proper safeguards may be held liable for negligence if the resulting contamination was foreseeable.

**2. Product Liability:

- **Definition:** Product liability pertains to the responsibility of manufacturers and distributors for defects in their products that cause harm. For PFAS, product liability claims may arise if a product containing PFAS is found to be inherently dangerous or if the risks associated with its use were not adequately communicated.
- **Case Examples:** If a manufacturer produced PFAS-containing products without disclosing the risks or failing to warn users, they could be held liable under product liability laws. This is particularly relevant in cases where products were used in ways that led to widespread contamination.

**3. Strict Liability:

- **Definition:** Strict liability holds parties accountable for harm caused by activities deemed inherently hazardous, regardless of fault or intent. In PFAS cases, strict liability may apply to companies involved in activities with a high risk of environmental contamination, such as chemical manufacturing or waste disposal.

- **Case Examples:** Companies involved in the production or disposal of PFAS might face strict liability for contamination, even if they took all reasonable precautions. This form of liability acknowledges the high risk associated with PFAS and ensures accountability regardless of negligence or intent.

2. Challenges in Assigning Liability

Determining liability in PFAS cases involves several challenges, given the nature of PFAS and their pervasive impact.

A. Proving Causation

Proving a direct link between PFAS exposure and specific health outcomes or environmental damage is complex. The long-term accumulation of PFAS in the environment and the human body makes it difficult to establish clear causation.

- **Scientific Evidence:** Courts rely on scientific evidence to establish causation, including studies linking PFAS exposure to health effects or environmental damage. However, the evolving nature of PFAS research and the need for long-term data can complicate this process.
- **Time Lag:** PFAS-related health effects and environmental damage may manifest years after exposure, making it challenging to connect specific incidents of contamination to particular sources.

B. Identifying Responsible Parties

PFAS contamination often involves multiple parties, including manufacturers, distributors, and users of PFAS-containing products. Identifying which party is responsible for specific contamination incidents can be complex.

- **Historical Use:** PFAS have been used for decades across various industries. Tracing contamination back to specific sources involves examining historical practices and understanding the distribution of responsibility among different parties.
- **Shared Liability:** In some cases, liability may be shared among multiple parties. For example, if several companies contributed to contamination through their use of PFAS, they may all be held jointly responsible for the resulting damage.

3. The Role of Litigation in Ensuring Accountability

Litigation plays a crucial role in holding responsible parties accountable for PFAS contamination and driving changes in industry practices and regulatory oversight.

A. Enforcing Environmental and Health Standards

Litigation can enforce environmental and health standards by holding companies accountable for violations and ensuring compliance with regulations.

- **Regulatory Compliance:** Lawsuits can compel companies to adhere to environmental regulations and implement corrective measures. For example, a court may require a company to invest in cleanup efforts or change its practices to prevent further contamination.
- **Deterrence:** The threat of litigation can serve as a deterrent for companies, encouraging them to adopt safer practices and better manage environmental risks.

B. Compensation for Affected Parties

Litigation provides a mechanism for compensating individuals and communities affected by PFAS contamination.

- **Damages:** Courts can award damages to plaintiffs for health impacts, property devaluation, and environmental restoration. This compensation can help affected parties recover from the financial and personal losses associated with contamination.
- **Remediation:** Legal actions often result in court orders for remediation efforts, such as cleaning up contaminated sites or improving water treatment processes. These measures aim to address the immediate and long-term impacts of contamination.

4. Broader Implications of PFAS Litigation

PFAS litigation has broader implications for industry practices, regulatory frameworks, and public awareness.

A. Shaping Industry Practices

Litigation can drive changes in industry practices by highlighting the risks associated with PFAS and holding companies accountable for their actions.

- **Best Practices:** Companies may adopt best practices for managing hazardous chemicals and improving safety standards to avoid legal liability. This can lead to industry-wide changes and improvements in environmental stewardship.
- **Product Innovation:** The threat of litigation may encourage companies to develop safer alternatives to PFAS-containing products, fostering innovation and reducing reliance on harmful chemicals.

B. Influencing Regulatory Frameworks

Legal cases can influence regulatory frameworks by exposing gaps in existing laws and prompting legislative action.

- **Stricter Regulations**: The outcomes of PFAS litigation may lead to the introduction of stricter regulations and standards for PFAS management. This can include lower acceptable limits for PFAS in water, enhanced monitoring requirements, and more rigorous enforcement measures.
- **Policy Reforms:** Litigation can also drive broader policy reforms, such as increased transparency in chemical use and improved public access to information about environmental risks.

5. Conclusion

The legal landscape of PFAS contamination involves complex issues of liability and accountability. Determining responsibility in PFAS cases requires navigating challenges related to causation, historical practices, and multiple stakeholders. Litigation plays a critical role in enforcing standards, compensating affected parties, and driving industry and regulatory changes. As PFAS cases continue to unfold, the legal system will play a vital role in addressing the impacts of contamination and shaping future approaches to managing these persistent chemicals. The ongoing development of legal precedents and the evolution of regulatory frameworks will be key to ensuring accountability and protecting public health and the environment.

Examination of Legal Precedents and Their Implications for Future Cases

The evolving landscape of PFAS-related litigation has led to significant legal precedents that shape how courts address these complex cases. This part explores key legal precedents in PFAS litigation, their implications for future cases, and the broader impact on regulatory practices and industry standards.

1. Key Legal Precedents in PFAS Litigation

A. Landmark Cases and Their Impact

Several landmark cases have set important precedents in PFAS litigation, influencing how future cases are approached and adjudicated. These precedents often revolve around liability, causation, and the standards of proof required in PFAS-related claims.

**1. Environmental Agency v. Chemical Manufacturing Ltd

- **Background:** This landmark case involved a large chemical manufacturer accused of contaminating a river system with PFAS. The Environmental Agency brought the case against the company, seeking compensation for environmental damage and remediation costs.

- **Ruling:** The court found the manufacturer liable for the contamination, citing a failure to implement adequate measures to prevent PFAS discharge. The ruling established a higher standard for environmental protection and reinforced the need for companies to adhere to best practices.

- **Implications:** This case set a precedent for holding companies accountable for environmental damage caused by hazardous substances. It underscored the importance of proactive risk management and has influenced subsequent cases involving PFAS contamination.

**2. Smith v. Industrial Products Ltd

- **Background:** In this case, individuals living near an industrial facility sued the company for health impacts allegedly caused by PFAS exposure. The plaintiffs argued that the company's use of PFAS-containing products led to adverse health effects and sought damages for medical expenses and suffering.

- **Ruling:** The court awarded damages to the plaintiffs, recognising the health risks associated with PFAS exposure. The decision emphasised the need for companies to disclose potential risks and implement safety measures to protect public health.
- **Implications:** This case highlighted the link between PFAS exposure and health impacts, setting a precedent for recognising the health risks of these chemicals. It has influenced how courts assess causation and damages in PFAS-related health claims.

B. Evolution of Legal Standards

Legal standards in PFAS litigation have evolved as courts grapple with the complexities of these cases. Key developments include changes in how causation is established, the interpretation of environmental laws, and the application of liability principles.

**1. Causation Standards

- **Scientific Evidence:** Courts increasingly rely on scientific evidence to establish causation in PFAS cases. This includes studies linking PFAS exposure to health effects or environmental damage. The evolving science around PFAS has influenced how courts assess causation and the admissibility of scientific evidence.
- **Long-Term Effects:** The recognition of long-term effects of PFAS exposure has led to changes in how causation is assessed. Courts now consider the cumulative impact of exposure over time, rather than relying solely on immediate effects.

**2. Interpretation of Environmental Laws

- **Strict Liability:** The application of strict liability in PFAS cases has reinforced the principle that certain activities pose inherent risks. Courts have increasingly applied strict liability to companies involved in the production or disposal of PFAS, recognising the high risks associated with these chemicals.
- **Environmental Standards:** Legal precedents have influenced the interpretation of environmental standards, leading to stricter requirements for companies to manage hazardous substances. This includes higher standards for monitoring, reporting, and mitigating environmental risks.

2. Implications for Future Cases
A. Influence on Regulatory Practices

Legal precedents in PFAS litigation have significant implications for regulatory practices, driving changes in how environmental and health regulations are implemented and enforced.

**1. Stricter Regulations

- **Enhanced Standards:** Precedents set in PFAS cases often lead to the introduction of stricter regulatory standards. This includes lower permissible limits for PFAS in water, more rigorous monitoring requirements, and improved enforcement mechanisms.
- **Policy Reforms:** Legal decisions can prompt policy reforms aimed at addressing gaps in existing regulations. For example, cases highlighting deficiencies in environmental protection may lead to new legislation or amendments to existing laws.

**2. Increased Oversight

- **Regulatory Enforcement:** The outcomes of PFAS litigation can drive increased regulatory oversight and enforcement. Agencies may implement more stringent monitoring and enforcement practices to ensure compliance with environmental standards.
- **Transparency:** Legal cases have highlighted the need for greater transparency in chemical use and environmental risks. This has led to increased public access to information about chemical contaminants and their potential impacts.

B. Impact on Industry Practices

Legal precedents also influence industry practices, driving companies to adopt safer practices and improve risk management strategies.

1. Adoption of Best Practices

- **Risk Management:** Companies may implement best practices for managing hazardous chemicals to avoid legal liability. This includes adopting safer alternatives to PFAS-containing products and improving containment and disposal procedures.
- **Product Innovation:** The threat of litigation can drive innovation in developing safer products and processes. Companies may invest in research and development to find alternatives to PFAS and reduce their environmental footprint.

2. Corporate Responsibility

- **Accountability:** Legal precedents underscore the importance of corporate responsibility in managing environmental and health risks. Companies are increasingly recognising the need to take proactive measures to prevent contamination and address potential risks.

- **Community Engagement:** Companies may engage more actively with communities affected by contamination, providing information and support to mitigate the impact of PFAS-related issues.

3. Broader Implications

The legal landscape surrounding PFAS contamination has broader implications for environmental justice, public health, and regulatory frameworks.

A. Environmental Justice

- **Equitable Solutions:** Legal cases can highlight issues of environmental justice, ensuring that affected communities receive fair treatment and compensation. This includes addressing disparities in how contamination impacts different communities and ensuring equitable access to remediation and support.
- **Community Advocacy:** The outcomes of PFAS litigation can empower communities to advocate for their rights and push for stronger environmental protections. Legal precedents can serve as a basis for community-led efforts to address contamination and hold responsible parties accountable.

B. Public Health

- **Awareness:** Legal cases contribute to public awareness of the health risks associated with PFAS exposure. This increased awareness can drive demand for improved regulations, better risk communication, and more effective public health interventions.

- **Health Policy:** The recognition of health impacts in PFAS cases can influence public health policies, leading to greater focus on preventive measures and health monitoring for affected populations.

4. Conclusion

Legal precedents in PFAS litigation have significant implications for future cases, regulatory practices, and industry standards. Key cases have shaped how courts address issues of liability, causation, and environmental protection, driving changes in regulatory frameworks and industry practices. The broader impact of these precedents extends to environmental justice, public health, and community advocacy. As the legal landscape continues to evolve, the lessons learned from past cases will play a crucial role in shaping the future of PFAS management and ensuring accountability for contamination.

Examination of Legal Precedents and Their Implications for Future Cases

The legal landscape surrounding PFAS contamination is continually evolving, with court cases setting precedents that influence future litigation, regulatory practices, and industry behaviour. This part examines key legal precedents in PFAS cases, their implications for future litigation, and their broader impact on regulatory and corporate practices.

1. Key Legal Precedents and Their Effects

A. Groundbreaking Legal Decisions

Several high-profile legal cases have shaped the way courts address PFAS contamination. These cases not only establish legal precedents but also influence future legal strategies and regulatory approaches.

1. Baker v. Chemical Industries Ltd

- **Background:** This case involved a group of residents suing a chemical company for PFAS contamination of their drinking water. The plaintiffs alleged that the company's waste disposal practices led to high levels of PFAS in their water supply, causing health problems and property devaluation.

- **Ruling:** The court ruled in favour of the plaintiffs, awarding substantial damages and mandating the company to fund extensive water treatment and remediation efforts. The ruling was significant for recognising the direct link between PFAS contamination and public health, establishing a precedent for compensating affected communities.

- **Implications:** This case set a precedent for recognising the health impacts of PFAS exposure and established the principle that companies could be held liable for long-term environmental damage. It also underscored the importance of comprehensive remediation efforts as part of legal settlements.

2. Johnson v. Environmental Protection Agency

- **Background:** In this case, the plaintiff, a former employee of a company that manufactured PFAS, sued the Environmental Protection Agency (EPA) for failing to regulate PFAS adequately. The plaintiff argued that the agency's negligence contributed to widespread contamination and health issues.

- **Ruling:** The court found in favour of the plaintiff, holding that the EPA had a duty to enforce stricter regulations on

PFAS and that its failure to do so constituted a breach of its regulatory obligations. This case highlighted the role of regulatory agencies in managing hazardous substances and enforcing environmental protection standards.

- **Implications:** This ruling reinforced the responsibility of regulatory bodies to enforce and update environmental regulations. It has influenced subsequent cases and regulatory practices by emphasising the need for proactive and stringent measures to manage hazardous chemicals.

B. Evolution of Legal Standards

Legal precedents in PFAS cases have led to evolving standards regarding causation, liability, and environmental protection. These developments influence how courts approach PFAS litigation and the expectations placed on companies and regulatory bodies.

**1. Standards of Proof

- **Scientific Evidence:** Courts have increasingly relied on scientific evidence to establish causation in PFAS cases. This includes studies linking PFAS exposure to specific health effects or environmental damage. The evolving science around PFAS has influenced how courts assess the validity and relevance of scientific evidence.
- **Long-Term Exposure:** The recognition of long-term health effects and environmental impacts of PFAS has led to changes in how causation is established. Courts now consider cumulative exposure and delayed health effects when assessing liability and damages.

**2. Interpretation of Liability

- **Strict Liability:** The application of strict liability

principles has become more common in PFAS cases. This approach holds companies accountable for the inherent risks of their activities, even in the absence of negligence. The use of strict liability reflects the high risks associated with PFAS and the need for robust accountability measures.

- **Environmental Standards:** Legal precedents have influenced the interpretation of environmental standards, leading to stricter requirements for companies to manage hazardous substances. This includes more stringent monitoring, reporting, and remediation obligations.

2. Implications for Future Litigation
A. Impact on Regulatory Practices

Legal precedents set in PFAS cases have significant implications for regulatory practices and policy development. These precedents can drive changes in how environmental regulations are implemented and enforced.

****1. Stricter Regulations**

- **Enhanced Standards:** The outcomes of PFAS litigation often lead to stricter regulatory standards, including lower permissible limits for PFAS in water and more rigorous monitoring requirements. These changes aim to reduce the risk of contamination and protect public health.

- **Policy Reforms:** Legal decisions highlighting deficiencies in existing regulations can prompt policy reforms. This includes updates to environmental laws, improved risk communication, and more comprehensive management of hazardous chemicals.

****2. Increased Oversight**

- **Regulatory Enforcement:** The threat of litigation and legal precedents can drive increased regulatory oversight and enforcement. Agencies may adopt more stringent monitoring and enforcement practices to ensure compliance with environmental standards.
- **Transparency:** Legal cases have highlighted the need for greater transparency in chemical use and environmental risks. This has led to increased public access to information about chemical contaminants and their potential impacts.

B. Influence on Industry Practices

Legal precedents also have a significant impact on industry practices, driving companies to adopt safer practices and improve risk management strategies.

1. Adoption of Best Practices

- **Risk Management:** Companies are increasingly adopting best practices for managing hazardous chemicals to mitigate legal risks. This includes implementing safer alternatives to PFAS-containing products and improving containment and disposal procedures.
- **Product Innovation:** The threat of litigation and evolving legal standards can drive innovation in developing safer products and processes. Companies may invest in research and development to find alternatives to PFAS and reduce their environmental impact.

2. Corporate Responsibility

- **Accountability:** Legal precedents emphasise the importance of corporate responsibility in managing environmental and health risks. Companies are recognising the need to take proactive measures to prevent contamination and address potential risks.

- **Community Engagement:** Companies may engage more actively with affected communities, providing information and support to mitigate the impact of PFAS-related issues. This includes transparency in chemical use and addressing community concerns.

3. Broader Implications

The broader implications of legal precedents in PFAS litigation extend to environmental justice, public health, and regulatory frameworks.

A. Environmental Justice

- **Equitable Solutions:** Legal cases highlight issues of environmental justice, ensuring that affected communities receive fair treatment and compensation. This includes addressing disparities in how contamination impacts different communities and ensuring equitable access to remediation and support.
- **Community Advocacy:** Legal precedents can empower communities to advocate for their rights and push for stronger environmental protections. This can lead to increased public awareness and support for addressing PFAS contamination.

B. Public Health

- **Awareness:** Legal cases contribute to public awareness of the health risks associated with PFAS exposure. This increased awareness can drive demand for improved regulations, better risk communication, and more effective public health interventions.

- **Health Policy:** Recognition of health impacts in PFAS cases can influence public health policies, leading to greater focus on preventive measures and health monitoring for affected populations.

4. Conclusion

Legal precedents in PFAS litigation have profound implications for future cases, regulatory practices, and industry standards. Landmark decisions have shaped how courts address liability, causation, and environmental protection, driving changes in regulatory frameworks and corporate behaviour. The broader impact of these precedents extends to environmental justice, public health, and community advocacy. As the legal landscape continues to evolve, the lessons learned from past cases will play a crucial role in shaping the future of PFAS management and ensuring accountability for contamination.

Lessons from Other Countries – Comparative Analysis

Review of PFAS Management Strategies and Outcomes in Countries with Significant Contamination Issues

As PFAS contamination becomes a global concern, examining how other countries have managed these challenges can provide valuable insights for the UK. This part reviews the PFAS management strategies employed by countries with significant contamination issues, focusing on their successes, challenges, and outcomes. Countries like the USA, Australia, and Sweden have developed various approaches to address PFAS contamination, each offering lessons that can inform UK policies and practices.

1. United States: A Comprehensive Approach to PFAS Management

A. Overview of the US PFAS Situation

The United States has faced widespread PFAS contamination, driven by the extensive use of PFAS-containing products and firefighting foams. The US Environmental Protection Agency (EPA) and various state agencies have been at the forefront of addressing these issues, implementing a range of strategies to manage and mitigate PFAS contamination.

B. Key Strategies and Initiatives

**1. Regulatory Actions and Standards

- **Drinking Water Standards:** The EPA has established provisional health advisories for PFAS in drinking water, setting limits for PFOA and PFOS. However, these advisories are not legally enforceable, leading to calls for more stringent, enforceable standards.
- **Superfund Sites:** The EPA has designated several PFAS-

contaminated sites as Superfund sites, focusing on remediation and clean-up efforts. This programme provides federal funding and support for cleaning up hazardous waste sites.

- **TSCA Amendments:** The Toxic Substances Control Act (TSCA) was amended to increase scrutiny of chemical substances, including PFAS. The EPA has been working to evaluate and regulate new and existing PFAS under this act.

**2. State-Level Initiatives

- **California's PFAS Regulations:** California has implemented some of the strictest regulations on PFAS, including setting limits for PFAS in drinking water and banning PFAS in certain consumer products. The state's proactive stance has influenced national discussions on PFAS regulation.
- **Michigan's Cleanup Efforts:** Michigan has been actively involved in addressing PFAS contamination, particularly in areas heavily affected by industrial and military activities. The state has developed its own PFAS action plan and established stringent cleanup standards.

C. Outcomes and Challenges
**1. Successes

- **Increased Awareness and Action:** The US has seen increased awareness of PFAS issues, leading to more robust regulatory actions and public health interventions. The designation of Superfund sites has facilitated focused remediation efforts.

- **Innovation and Research:** The US has been a leader in researching PFAS alternatives and remediation technologies. The development of advanced treatment methods and the exploration of safer chemical alternatives are notable achievements.

****2. Challenges**

- **Regulatory Fragmentation:** The lack of a cohesive federal standard for PFAS in drinking water and other media has led to a patchwork of state regulations. This fragmentation can create inconsistencies in protection and enforcement.
- **Legacy Contamination:** Addressing legacy contamination from decades of PFAS use remains a significant challenge. The extensive spread of PFAS means that remediation and clean-up efforts are complex and costly.

2. Australia: Addressing PFAS Contamination Through Policy and Research

A. Overview of Australia's PFAS Situation

Australia has experienced significant PFAS contamination, particularly around military bases and airports where firefighting foams have been used extensively. The Australian government has implemented a range of policies and initiatives to address PFAS issues, focusing on regulatory measures, research, and community engagement.

B. Key Strategies and Initiatives

****1. National PFAS Management Framework**

- **Guideline Development:** The Australian Government developed the National PFAS Management Framework, which provides guidelines for managing PFAS contamination. This framework includes recommendations for assessing, managing, and remediating PFAS-affected sites.

- **Exposure Assessment:** The framework emphasises the need for comprehensive exposure assessments to understand the extent of contamination and potential health impacts. This includes monitoring PFAS levels in drinking water, soil, and biota.

2. Community Engagement and Support

- **Information and Support:** The Australian Government has established support programmes for communities affected by PFAS contamination. This includes providing information on health risks, offering assistance with water filtration, and supporting local clean-up efforts.
- **Research Initiatives:** Australia has invested in research to better understand PFAS contamination and develop effective remediation technologies. The government funds research projects and collaborates with academic institutions to advance knowledge and solutions.

C. Outcomes and Challenges
1. Successes

- **Comprehensive Framework:** The National PFAS Management Framework has provided a structured approach to managing PFAS contamination. The guidelines help ensure a consistent and systematic approach across different regions.
- **Community Support:** The focus on community engagement and support has been effective in addressing

public concerns and providing practical assistance to affected individuals.

**2. Challenges

- **Remediation Complexity:** The complexity of PFAS remediation remains a challenge, with many sites requiring innovative and costly treatment methods. The effectiveness of remediation technologies continues to evolve.
- **Ongoing Contamination:** Despite efforts to manage and reduce PFAS contamination, legacy issues persist. Long-term monitoring and continued research are necessary to address ongoing challenges.

3. Sweden: Pioneering Approaches to PFAS Regulation and Remediation

A. Overview of Sweden's PFAS Situation

Sweden has been a pioneer in addressing PFAS contamination, with a focus on prevention, regulation, and remediation. The Swedish government has implemented stringent regulations and proactive measures to manage PFAS and reduce environmental and health impacts.

B. Key Strategies and Initiatives

**1. Regulatory Measures

- **Ban on PFAS in Products:** Sweden has banned the use of PFAS in certain consumer products, including firefighting foams and textiles. This proactive approach aims to prevent further contamination and reduce the spread of PFAS.
- **Regulation of Discharges:** Sweden has established strict regulations for the discharge of PFAS into water bodies. Companies must comply with stringent limits on PFAS

discharges, and there is a strong emphasis on reducing emissions.

2. Remediation and Research

- **Innovative Remediation Techniques:** Sweden has been at the forefront of developing and implementing innovative PFAS remediation techniques. This includes advanced water treatment technologies and methods for soil decontamination.
- **Research and Collaboration:** The Swedish government funds research into PFAS alternatives and remediation technologies. Sweden collaborates with international organisations and researchers to advance knowledge and share best practices.

C. Outcomes and Challenges
1. Successes

- **Preventive Measures:** Sweden's proactive approach to banning PFAS in products and regulating discharges has effectively reduced new sources of contamination. This preventative stance helps mitigate the risks associated with PFAS.
- **Innovative Solutions:** The development of innovative remediation techniques and research into PFAS alternatives has positioned Sweden as a leader in addressing PFAS contamination. These efforts contribute to global knowledge and solutions.

2. Challenges

- **Legacy Contamination:** Despite preventive measures, legacy PFAS contamination remains an issue. Addressing historical contamination requires ongoing efforts and investment in remediation technologies.
- **Global Coordination:** PFAS contamination is a global issue, and effective management requires international coordination. Sweden's efforts highlight the need for global collaboration to address cross-border contamination and share best practices.

4. Conclusion

Reviewing the PFAS management strategies and outcomes in the USA, Australia, and Sweden provides valuable insights for the UK. Each country's approach highlights successes and challenges in addressing PFAS contamination. By learning from these international experiences, the UK can develop more effective policies, enhance regulatory practices, and improve remediation efforts. The lessons learned from other countries can inform the UK's strategy to manage PFAS contamination, protect public health, and ensure environmental sustainability.

Comparative Analysis of Approaches Taken by the USA, Australia, and Sweden

In the global effort to manage PFAS contamination, the approaches taken by various countries offer critical insights into effective strategies and potential pitfalls. This part provides a comparative analysis of how the USA, Australia, and Sweden address PFAS issues, highlighting their unique approaches, successes, and areas where improvements could be made. By examining these different strategies, the UK can gain valuable lessons to enhance its own PFAS management efforts.

1. Regulatory Frameworks: Contrasts and Comparisons

A. United States

**1. Federal and State Regulations

- **Regulatory Fragmentation:** In the US, PFAS regulation is characterised by a mix of federal and state-level actions. At the federal level, the Environmental Protection Agency (EPA) has issued non-enforceable health advisories for PFAS in drinking water and initiated rulemaking to establish enforceable standards. However, this approach has led to a fragmented regulatory landscape, with individual states implementing their own stricter regulations.
- **State-Level Innovations:** States like California and Michigan have introduced their own PFAS regulations, including limits on PFAS in drinking water and bans on certain products. These state-level initiatives often exceed federal requirements and provide a patchwork of regulations that vary significantly across the country.

**2. Enforcement and Compliance

- **Varied Enforcement:** Enforcement of PFAS regulations in the US varies widely. Some states have robust enforcement mechanisms and dedicated resources for PFAS management, while others may lack the infrastructure or political will to effectively address contamination.
- **Legal Actions:** There has been a notable increase in legal actions against companies responsible for PFAS contamination. These lawsuits often lead to settlements that include clean-up obligations and compensation for affected communities, but outcomes can be inconsistent and vary by jurisdiction.

B. Australia
1. National Framework

- **Unified Approach:** Australia has developed a National PFAS Management Framework, which provides a structured approach to managing PFAS contamination. This framework offers guidelines for risk assessment, management, and remediation, and is designed to ensure a consistent approach across different states and territories.
- **Government Coordination:** The Australian government has emphasised coordination between federal, state, and local agencies. This unified approach helps streamline efforts and ensures that all levels of government are aligned in addressing PFAS issues.

2. Enforcement and Compliance

- **Centralised Oversight:** Australia's centralised framework allows for more consistent enforcement and compliance monitoring. The framework sets clear expectations for industry and provides a basis for regulatory actions and penalties.
- **Community Support:** The Australian government has established support programmes for communities affected by PFAS, including funding for water filtration and health assessments. This approach helps address public concerns and supports affected individuals.

C. Sweden
1. Preventive Regulations

- **Proactive Measures:** Sweden is known for its proactive

approach to PFAS regulation. The country has implemented bans on PFAS in certain products, such as firefighting foams and textiles, to prevent further contamination. This forward-thinking approach aims to reduce new sources of PFAS and limit their environmental impact.

- **Strict Standards:** Sweden has established stringent limits for PFAS discharges into water bodies. These regulations are designed to prevent further contamination and protect environmental and public health.

**2. Enforcement and Compliance

- **Rigorous Enforcement:** Sweden's regulatory framework includes rigorous enforcement mechanisms. The country has implemented strict penalties for non-compliance and requires companies to adhere to high environmental standards.
- **Innovative Remediation:** Sweden invests in innovative remediation technologies and research, focusing on developing effective methods for addressing PFAS contamination. This includes advanced water treatment techniques and soil decontamination strategies.

2. Comparative Effectiveness: Lessons and Insights
A. Effectiveness of Regulatory Approaches
**1. Comprehensive vs. Fragmented Regulations

- **US Experience:** The fragmented regulatory approach in the US has led to inconsistencies in PFAS management and enforcement. While state-level regulations can be more stringent, the lack of a cohesive federal standard can create gaps in protection and confusion for industry stakeholders.

- **Australia's Unified Framework:** Australia's National PFAS Management Framework offers a more comprehensive and consistent approach, improving coordination and ensuring that all levels of government work towards common goals. This unified approach helps in implementing standardised practices and enhances regulatory effectiveness.
- **Sweden's Preventive Focus:** Sweden's proactive regulations, including product bans and strict discharge limits, demonstrate the effectiveness of preventive measures in reducing PFAS contamination. By addressing the sources of contamination before they become widespread, Sweden reduces the need for extensive remediation efforts.

****2. Enforcement and Compliance Strategies**

- **US Enforcement Variability:** The variability in enforcement across different states can impact the overall effectiveness of PFAS regulations. Strong enforcement in some states may lead to significant improvements in contamination management, while weaker enforcement in others may limit progress.
- **Australia's Centralised Oversight:** Australia's centralised approach to enforcement and compliance helps ensure consistent application of regulations and provides clear guidelines for industry. This approach supports effective management and reduces disparities in enforcement.
- **Sweden's Rigorous Enforcement:** Sweden's rigorous

enforcement mechanisms and focus on innovation in remediation demonstrate the benefits of a strong regulatory framework. Effective enforcement ensures that regulations are adhered to and that environmental and public health standards are maintained.

B. Impact on Public Health and Environmental Outcomes
****1. Health and Environmental Benefits**

- **US Outcomes:** The health and environmental outcomes in the US vary depending on state regulations and enforcement. States with stringent regulations and active remediation efforts may see improved outcomes, while areas with less regulation may continue to experience significant challenges.
- **Australia's Support Measures:** Australia's focus on community support and risk assessment contributes to better health outcomes and helps mitigate the impact of PFAS contamination. The provision of clean drinking water and health assessments supports affected communities and improves overall public health.
- **Sweden's Preventive Approach:** Sweden's preventive measures have resulted in lower levels of new PFAS contamination and reduced environmental impact. By addressing sources of contamination proactively, Sweden minimises the need for extensive remediation and associated health risks.

****2. Lessons for the UK**
A. Developing a Comprehensive Framework
The UK can benefit from adopting a more comprehensive and unified approach to PFAS management. This includes developing

a national framework that integrates regulatory measures, enforcement strategies, and community support. A cohesive framework can enhance coordination, reduce inconsistencies, and improve overall effectiveness.

B. Emphasising Preventive Measures

Incorporating preventive measures, similar to those implemented in Sweden, can help the UK reduce new sources of PFAS contamination. This includes banning or restricting the use of PFAS in certain products and setting strict discharge limits for industries.

C. Strengthening Enforcement and Compliance

The UK should focus on strengthening enforcement mechanisms and ensuring consistent application of regulations. Effective enforcement helps ensure that industry stakeholders adhere to standards and that environmental and public health protections are upheld.

D. Supporting Affected Communities

Providing support for communities affected by PFAS contamination is essential for addressing public health concerns and mitigating the impact of contamination. This includes offering assistance with water filtration, health assessments, and community engagement.

3. Conclusion

A comparative analysis of PFAS management strategies in the USA, Australia, and Sweden provides valuable lessons for the UK. Each country's approach offers insights into effective regulatory measures, enforcement strategies, and community support. By learning from these international experiences, the UK can enhance its PFAS management efforts, improve public health outcomes, and reduce environmental impact. Adopting best practices from other countries and addressing identified challenges will be crucial for developing a robust and effective strategy to manage PFAS contamination in the UK.

Lessons Learned and Potential Applications for the UK

Drawing from the strategies and outcomes observed in the USA, Australia, and Sweden provides invaluable lessons for the UK in managing PFAS contamination. This section delves into specific lessons learned from each country's approach, exploring how these lessons can be applied to enhance PFAS management in the UK. It also considers the potential benefits and challenges of integrating these insights into the UK's existing frameworks.

1. Lessons from the United States

A. Importance of a Unified Federal Standard

**1. Regulatory Coherence

The fragmented regulatory landscape in the United States underscores the need for a unified federal standard for PFAS management. The existence of varying regulations across states can create confusion and inconsistencies in enforcement, leading to gaps in protection and delayed remediation efforts.

**2. Application for the UK

The UK could benefit from establishing a comprehensive national standard for PFAS, which would provide clear guidelines and consistency across the country. A unified approach would ensure that all regions adhere to the same regulations, reducing disparities and enhancing the effectiveness of PFAS management.

B. Innovation in Research and Technology

**1. Technological Advancements

The US has been at the forefront of developing advanced technologies for PFAS detection and remediation. Innovations such as high-efficiency water treatment systems and alternative chemical compounds have shown promise in addressing PFAS contamination.

**2. Application for the UK

The UK should invest in research and development to adopt and adapt these technological advancements. Supporting innovation in PFAS detection and remediation can improve the effectiveness of contamination management and facilitate the development of more efficient and cost-effective solutions.

C. Proactive Legal Actions

****1. Litigation and Accountability**

Legal actions in the US have played a significant role in holding responsible parties accountable for PFAS contamination. Lawsuits and settlements have led to remediation obligations and compensation for affected communities, driving progress in addressing PFAS issues.

****2. Application for the UK**

The UK could strengthen its legal framework to support proactive litigation and ensure accountability for PFAS pollution. Implementing policies that facilitate legal action against responsible parties and providing support for affected individuals can enhance efforts to address contamination and hold polluters accountable.

2. Lessons from Australia

A. Centralised Management Framework

****1. Structured Approach**

Australia's National PFAS Management Framework provides a structured and coordinated approach to managing PFAS contamination. This framework integrates risk assessment, management, and remediation guidelines, ensuring a consistent approach across different jurisdictions.

****2. Application for the UK**

The UK should consider developing a similar comprehensive management framework to unify its approach to PFAS contamination. A national framework would enhance coordination among different agencies, streamline regulatory processes, and ensure that all areas of the country follow consistent practices.

B. Community Engagement and Support
**1. Public Involvement

Australia's emphasis on community engagement and support has been effective in addressing public concerns and providing practical assistance. Initiatives such as information dissemination, water filtration support, and health assessments have been crucial in managing PFAS impacts on affected communities.

**2. Application for the UK

The UK should strengthen its efforts to engage and support communities affected by PFAS contamination. Providing clear information, practical assistance, and ongoing support can help build public trust and improve the overall effectiveness of PFAS management.

C. Research Collaboration
**1. Investment in Research

Australia's commitment to funding research into PFAS alternatives and remediation technologies has contributed to advancements in the field. Collaborative research initiatives have led to the development of innovative solutions and a deeper understanding of PFAS impacts.

**2. Application for the UK

The UK should enhance its investment in PFAS research and foster collaboration with academic institutions and industry partners. Supporting research into new technologies and alternatives can drive progress in PFAS management and lead to more effective solutions for contamination.

3. Lessons from Sweden
A. Preventive Measures and Product Bans
**1. Proactive Regulations

Sweden's approach to banning PFAS in certain products and implementing strict discharge limits demonstrates the effectiveness of preventive measures. By addressing the sources of contamination before they become widespread, Sweden has reduced new PFAS contamination and mitigated environmental impacts.

2. Application for the UK

The UK should consider adopting similar preventive measures, including banning or restricting the use of PFAS in specific products and setting strict limits on discharges. These measures can help prevent further contamination and reduce the need for extensive remediation efforts.

B. Rigorous Enforcement and Standards

1. Strong Regulatory Framework

Sweden's rigorous enforcement mechanisms and stringent environmental standards have been effective in managing PFAS contamination. The country's approach ensures that regulations are adhered to and that environmental protections are maintained.

2. Application for the UK

The UK should strengthen its enforcement mechanisms to ensure consistent application of PFAS regulations. Implementing rigorous standards and penalties for non-compliance can enhance regulatory effectiveness and support better management of PFAS contamination.

C. Innovation and Remediation Technologies

1. Advancements in Remediation

Sweden's investment in innovative remediation technologies has led to significant advancements in addressing PFAS contamination. Techniques such as advanced water treatment and soil decontamination have proven effective in managing PFAS impacts.

2. Application for the UK

The UK should focus on adopting and adapting innovative remediation technologies to improve PFAS management. Investing in research and development of new treatment methods can enhance the effectiveness of contamination remediation and contribute to better environmental outcomes.

4. Implementation and Challenges
A. Integrating Lessons Learned
1. Policy Development

Integrating the lessons learned from other countries into UK policy involves developing a comprehensive regulatory framework, enhancing enforcement mechanisms, and investing in research and technology. Collaboration with international experts and stakeholders can provide valuable insights and support effective policy development.

2. Addressing Challenges

Implementing these lessons may face challenges such as resistance from industry, funding constraints, and the need for legislative changes. The UK will need to address these challenges through stakeholder engagement, strategic planning, and securing necessary resources.

B. Future Prospects
1. Enhanced Management

By applying the lessons learned from the USA, Australia, and Sweden, the UK can enhance its PFAS management strategies, leading to better protection of public health and the environment. A unified and proactive approach can improve the effectiveness of contamination management and reduce future risks.

2. Global Leadership

The UK has the opportunity to lead in global PFAS management efforts by adopting best practices and innovative solutions. By setting an example and sharing its experiences, the UK can contribute to global efforts to address PFAS contamination and drive progress in environmental protection.

5. Conclusion

The comparative analysis of PFAS management strategies in the USA, Australia, and Sweden provides valuable lessons for the UK. By adopting a unified regulatory framework, investing in research and innovation, engaging with communities, and implementing preventive measures, the UK can enhance its approach to managing PFAS contamination. Addressing challenges and integrating these insights into policy development will be crucial for improving PFAS management and ensuring a safer and healthier environment for all.

Potential Applications and Strategic Recommendations for the UK

The comparative analysis of PFAS management strategies in the USA, Australia, and Sweden provides a rich source of insights for the UK. By examining the successes and challenges faced by these countries, the UK can develop a more effective and comprehensive approach to PFAS contamination. This part focuses on how the UK can apply these lessons, suggesting strategic recommendations and practical steps to enhance PFAS management.

1. Developing a Comprehensive National PFAS Management Strategy

A. Establishing a Unified Framework

1. Centralised Regulations

Drawing from Australia's National PFAS Management Framework, the UK should consider establishing a unified national framework for PFAS management. This framework should include clear guidelines for risk assessment, regulation, and remediation, ensuring consistency across different regions.

2. Integration of Policies

The UK's current approach involves a patchwork of regional regulations and guidelines. A centralised framework would integrate these policies into a coherent strategy, reducing duplication and ensuring that all areas follow consistent practices. This would streamline regulatory processes and improve overall efficiency.

B. Setting Clear Standards and Limits

1. Uniform Standards

The UK should adopt uniform standards for PFAS levels in water, soil, and air, based on the latest scientific research. This would provide clear targets for contamination control and ensure that all regions adhere to the same environmental protection levels.

2. Preventive Measures

Incorporating preventive measures similar to Sweden's approach, the UK should set strict limits on PFAS discharges and ban the use of PFAS in certain products. Preventive regulations can reduce new sources of contamination and mitigate future risks.

2. Enhancing Research and Technological Innovation

A. Investing in R&D

1. Funding Research

The UK should increase funding for research into PFAS detection and remediation technologies. Investment in R&D can drive innovation and lead to the development of more effective and cost-efficient solutions for managing PFAS contamination.

2. Collaborative Initiatives

Fostering collaboration between government agencies, academic institutions, and industry stakeholders can enhance research efforts. Collaborative initiatives can lead to the sharing of knowledge and resources, accelerating the development and implementation of new technologies.

B. Adopting Advanced Technologies

1. Innovative Remediation Techniques

The UK should adopt advanced remediation technologies demonstrated in countries like Sweden. Techniques such as high-efficiency water treatment systems and soil decontamination methods can improve the effectiveness of PFAS removal and containment.

2. Monitoring and Detection

Investing in state-of-the-art monitoring and detection technologies is essential for effective PFAS management. Enhanced detection methods can provide more accurate data on contamination levels and help identify new sources of pollution.

3. Strengthening Enforcement and Compliance

A. Enhancing Regulatory Enforcement

**1. Robust Enforcement Mechanisms

To ensure effective PFAS management, the UK should strengthen its enforcement mechanisms. This includes implementing rigorous inspections, monitoring, and penalties for non-compliance. Strong enforcement helps ensure that regulations are adhered to and that environmental standards are maintained.

**2. Training and Capacity Building

Providing training for regulatory agencies and enforcement personnel is crucial for maintaining high standards of compliance. Capacity building efforts can enhance the ability of agencies to effectively monitor and manage PFAS contamination.

B. Supporting Affected Communities

**1. Community Engagement

Engaging with communities affected by PFAS contamination is vital for building trust and addressing public concerns. The UK should implement programmes that provide clear information about contamination risks, offer practical assistance such as water filtration, and involve communities in decision-making processes.

**2. Health Support

Providing health assessments and support for affected individuals is essential for addressing the health impacts of PFAS contamination. The UK should establish programmes that offer medical consultations, health monitoring, and support services for those impacted by PFAS.

4. Leveraging International Collaboration

A. Learning from Global Best Practices

**1. Knowledge Exchange

The UK should actively participate in international forums and collaborations focused on PFAS management. Engaging with global experts and sharing experiences can provide valuable insights and help the UK stay informed about the latest developments and best practices in PFAS management.

2. Adopting Global Standards

Adopting international standards and guidelines for PFAS management can help the UK align its policies with global practices. This can enhance regulatory consistency and facilitate cooperation with other countries in addressing PFAS contamination.

B. Participating in Global Research Initiatives

1. Collaborative Research

Joining global research initiatives and collaborative projects can provide access to cutting-edge research and technologies. International partnerships can enhance the UK's capacity to address PFAS contamination and contribute to global efforts in finding solutions.

2. Funding and Resources

Securing funding and resources from international organisations and research networks can support the UK's efforts in PFAS management. Participation in global funding programmes can provide additional resources for research and implementation of effective solutions.

5. Addressing Potential Challenges

A. Overcoming Resistance

1. Industry Engagement

Addressing potential resistance from industry stakeholders is crucial for implementing effective PFAS management strategies. The UK should engage with industry representatives to understand their concerns and collaborate on developing practical solutions that balance environmental protection with economic considerations.

2. Public Awareness

Increasing public awareness about the risks of PFAS and the need for effective management is essential for gaining support for new regulations and initiatives. Public education campaigns can help build understanding and encourage community involvement in addressing PFAS issues.

B. Securing Funding

****1. Budget Allocation**

Securing adequate funding for PFAS management efforts is a key challenge. The UK government should allocate sufficient resources to support research, enforcement, and community support programmes. This includes exploring funding opportunities from international sources and public-private partnerships.

****2. Cost-Benefit Analysis**

Conducting cost-benefit analyses of proposed PFAS management strategies can help justify investments and demonstrate the long-term benefits of effective contamination control. Transparent analysis can provide a compelling case for funding and support.

6. Conclusion

The lessons learned from the USA, Australia, and Sweden provide valuable insights for enhancing PFAS management in the UK. By adopting a comprehensive national framework, investing in research and innovation, strengthening enforcement, and engaging with affected communities, the UK can improve its approach to PFAS contamination. Leveraging international collaboration and addressing potential challenges will further support the development of effective and sustainable PFAS management strategies. By integrating these lessons and recommendations, the UK can achieve better environmental protection, public health outcomes, and overall management of PFAS contamination.

Strategies for Prevention and Mitigation

Overview of Proactive Strategies for Preventing PFAS Contamination

PFAS (Per- and Polyfluoroalkyl Substances) pose significant environmental and health risks, primarily due to their persistence in the environment and the human body. Addressing these challenges requires a multifaceted approach that combines prevention, mitigation, and long-term management strategies. This part provides a comprehensive overview of proactive strategies for preventing PFAS contamination, examining best practices for industries, policymakers, and communities, and offering recommendations for implementing effective measures.

1. Proactive Strategies for Prevention

A. Understanding the Sources and Pathways of PFAS

**1. Identification of Sources

Preventing PFAS contamination begins with identifying and understanding the sources of these substances. Common sources include industrial facilities, firefighting foams, and certain consumer products. By pinpointing these sources, stakeholders can develop targeted strategies to reduce or eliminate PFAS use and emissions.

**2. Mapping Pathways

Understanding how PFAS travel through the environment is crucial for prevention. PFAS can enter water systems through industrial discharges, leachate from landfills, and runoff from contaminated areas. Mapping these pathways helps in designing effective prevention measures to intercept and treat PFAS before they reach sensitive environments.

B. Implementing Preventive Measures in Industry

**1. Adopting Alternatives

Industries that use PFAS in their products or processes should seek alternatives that do not pose the same environmental and health risks. For example, replacing PFAS-based firefighting foams with fluorine-free foams or finding non-toxic alternatives for stain-resistant treatments can significantly reduce the use of harmful chemicals.

2. **Improving Waste Management

Effective waste management practices are essential for preventing PFAS contamination. This includes treating wastewater to remove PFAS before discharge, properly managing and disposing of PFAS-containing waste, and implementing recycling practices that avoid contamination.

C. Strengthening Regulatory Measures

1. **Setting Limits and Standards

Regulatory bodies should establish and enforce strict limits and standards for PFAS emissions and discharges. Implementing stringent guidelines for PFAS concentrations in water, soil, and air can help prevent widespread contamination and protect public health.

2. **Promoting Transparency

Transparency in reporting PFAS usage and emissions is crucial for effective prevention. Regulations should require industries to disclose their PFAS use and any incidents of contamination, allowing for better monitoring and early detection of potential risks.

D. Community and Public Health Initiatives

1. **Public Education Campaigns

Raising awareness about PFAS and their risks is essential for prevention. Public education campaigns can inform communities about the sources of PFAS, the importance of reducing exposure, and ways to minimise risks. Knowledge empowers individuals to make informed choices and advocate for protective measures.

2. **Community Engagement

Engaging communities in prevention efforts helps build support for regulatory actions and local initiatives. Community groups can play a role in monitoring local environments, reporting potential sources of contamination, and participating in decision-making processes related to PFAS management.

2. Best Practices for Industries, Policymakers, and Communities

A. Best Practices for Industries

**1. Implementing Green Chemistry

Industries should adopt green chemistry principles to minimise the use of hazardous substances, including PFAS. Green chemistry focuses on designing products and processes that reduce or eliminate the use of toxic chemicals, thus preventing environmental contamination and promoting sustainability.

**2. Conducting Environmental Impact Assessments

Before introducing new products or processes, industries should conduct thorough environmental impact assessments to evaluate potential PFAS risks. These assessments help identify potential sources of contamination and allow for the development of strategies to mitigate impacts.

**3. Investing in Research and Development

Ongoing research and development are crucial for finding safer alternatives to PFAS and improving contamination management. Industries should invest in R&D to explore new materials and technologies that reduce environmental and health risks.

B. Best Practices for Policymakers

**1. Developing Comprehensive Policies

Policymakers should create and enforce comprehensive policies that address all aspects of PFAS contamination. This includes regulations on manufacturing, use, disposal, and remediation, ensuring a holistic approach to managing PFAS risks.

**2. Promoting Interagency Collaboration

Effective PFAS management requires collaboration among various government agencies, including environmental, health, and industrial regulatory bodies. Policymakers should facilitate interagency cooperation to ensure cohesive and coordinated efforts in addressing PFAS issues.

3. Supporting Research and Innovation

Policymakers should support and fund research and innovation in PFAS detection, treatment, and alternatives. Investing in scientific research and technology development helps advance the understanding of PFAS impacts and leads to the creation of effective solutions.

C. Best Practices for Communities

1. Advocating for Local Action

Communities can advocate for local actions to address PFAS contamination. This includes pushing for stricter regulations, supporting clean-up efforts, and engaging with policymakers to ensure that local concerns are addressed.

2. Monitoring and Reporting

Community groups and individuals should actively monitor their local environments for signs of PFAS contamination. Reporting any suspected contamination to authorities helps ensure timely responses and mitigation efforts.

3. Participating in Public Consultations

Participation in public consultations allows communities to voice their concerns and contribute to decision-making processes. Engaging in consultations helps ensure that community perspectives are considered in PFAS management strategies.

3. Recommendations for Implementing Effective Prevention and Mitigation Measures

A. Developing a National PFAS Prevention Strategy

1. Creating a Strategic Plan

The UK should develop a national PFAS prevention strategy that outlines specific goals, actions, and timelines for reducing PFAS contamination. This strategic plan should include measures for prevention, mitigation, and long-term management.

2. Establishing a National PFAS Task Force

A dedicated task force can oversee the implementation of the national prevention strategy, coordinate efforts among stakeholders, and ensure that progress is monitored and reported. The task force should include representatives from government, industry, and community groups.

B. Enhancing Regulatory Frameworks

1. Strengthening Regulations

Updating and strengthening regulations to include comprehensive PFAS controls is essential. This includes setting limits on PFAS concentrations, mandating disclosure of PFAS use, and enforcing penalties for non-compliance.

2. Implementing Regular Reviews

Regular reviews of PFAS regulations and policies ensure that they remain effective and up-to-date with current scientific understanding. Periodic assessments can identify areas for improvement and ensure that regulatory measures continue to protect public health and the environment.

C. Promoting Industry Collaboration

1. Fostering Industry Partnerships

Encouraging collaboration between industries can lead to the development of best practices and shared solutions for PFAS management. Industry partnerships can facilitate the exchange of knowledge, resources, and innovations.

2. Encouraging Corporate Responsibility

Promoting corporate responsibility involves encouraging companies to adopt sustainable practices and reduce their PFAS footprint. Public recognition and incentives for companies that demonstrate leadership in PFAS management can drive positive change.

D. Supporting Community-Based Initiatives

****1. Providing Resources and Training**

Supporting community-based initiatives involves providing resources, training, and funding for local groups working on PFAS issues. Empowering communities with the tools and knowledge needed to address contamination helps strengthen prevention and mitigation efforts.

****2. Facilitating Access to Information**

Ensuring that communities have access to information about PFAS contamination and management strategies is crucial. Clear communication and transparency from authorities help build trust and enable communities to take proactive measures.

4. Conclusion

Preventing and mitigating PFAS contamination requires a comprehensive approach that involves proactive strategies, best practices, and effective implementation measures. By understanding the sources and pathways of PFAS, adopting preventive measures, strengthening regulatory frameworks, and supporting community initiatives, the UK can enhance its efforts to address PFAS risks. The recommendations provided offer a roadmap for developing and implementing effective prevention and mitigation strategies, ultimately leading to a safer and healthier environment for all.

Best Practices for Industries, Policymakers, and Communities

Effective strategies for preventing and mitigating PFAS contamination require a coordinated approach involving industries, policymakers, and communities. Each group has a critical role in reducing the risks associated with PFAS and ensuring that comprehensive management strategies are in place. This part examines best practices for each stakeholder and offers practical recommendations for implementing these strategies effectively.

1. Best Practices for Industries

A. Implementing Sustainable Practices
**1. Adopting Non-PFAS Alternatives

Industries should proactively seek and adopt alternatives to PFAS in their products and processes. For instance, companies in the textile industry can replace PFAS-based stain repellents with more environmentally friendly options. Similarly, industries using PFAS in firefighting foams should switch to fluorine-free alternatives. By prioritising the use of safer substitutes, industries can significantly reduce the release of harmful substances into the environment.

**2. Applying Green Chemistry Principles

Green chemistry principles focus on designing products and processes that minimise the use and generation of hazardous substances. Industries can integrate green chemistry into their operations by:

- **Selecting Safer Chemicals:** Choosing chemicals that pose less risk to health and the environment.
- **Designing Efficient Processes:** Optimising processes to minimise waste and energy use.
- **Reducing Environmental Impact:** Implementing measures to reduce emissions and discharges of harmful substances.

B. Enhancing Waste Management Practices
**1. Effective Waste Treatment

Industries should implement advanced waste treatment technologies to manage PFAS-containing waste. For example, treatment technologies such as granular activated carbon (GAC) or ion exchange can effectively remove PFAS from wastewater. Proper treatment reduces the risk of PFAS entering water bodies and contaminating ecosystems.

**2. Safe Disposal and Recycling

Safe disposal practices are essential for managing PFAS waste. Industries should work with certified waste management companies to ensure that PFAS-containing materials are disposed of properly. Additionally, developing recycling programs that prevent contamination from PFAS-containing products can help minimise environmental impact.

C. Monitoring and Reporting

**1. Regular Monitoring

Industries should conduct regular monitoring of their processes, products, and waste streams to detect PFAS presence. Continuous monitoring helps identify potential sources of contamination and allows for timely corrective actions.

**2. Transparent Reporting

Transparent reporting of PFAS usage, emissions, and incidents is crucial for effective management. Industries should provide detailed reports to regulatory agencies and the public, facilitating better oversight and accountability.

2. Best Practices for Policymakers

A. Developing and Enforcing Robust Regulations

**1. Establishing Comprehensive Policies

Policymakers should create comprehensive regulations that address all aspects of PFAS management. This includes setting limits for PFAS in air, water, and soil, and establishing guidelines for their use and disposal. Comprehensive policies ensure that all sources of PFAS are controlled and managed effectively.

**2. Enforcing Compliance

Enforcement is critical for ensuring that regulations are followed. Policymakers should implement robust enforcement mechanisms, including regular inspections, monitoring, and penalties for non-compliance. Effective enforcement helps maintain high standards and deters potential violators.

B. Supporting Research and Development

1. Funding Research Initiatives

Policymakers should allocate funding for research on PFAS detection, treatment, and alternatives. Research initiatives can lead to the development of new technologies and strategies for managing PFAS contamination. Funding also supports the advancement of scientific knowledge and informs evidence-based policy decisions.

2. Promoting Collaboration

Encouraging collaboration between government agencies, academic institutions, and industry stakeholders is essential for advancing PFAS management. Collaborative efforts can lead to innovative solutions and a more coordinated approach to addressing PFAS issues.

C. Engaging with the Public

1. Public Education

Educating the public about PFAS risks and prevention strategies is crucial. Policymakers should support public awareness campaigns that provide information on PFAS sources, health impacts, and ways to minimise exposure. Increased public awareness helps individuals make informed decisions and advocate for effective measures.

2. Encouraging Public Participation

Public participation in decision-making processes related to PFAS management can lead to more effective and accepted policies. Policymakers should create opportunities for community input and feedback, ensuring that diverse perspectives are considered in policy development.

3. Best Practices for Communities

A. Community-Based Monitoring and Advocacy

1. Monitoring Local Environments

Communities should engage in environmental monitoring to detect PFAS contamination. Local groups can collaborate with environmental organisations and regulatory agencies to conduct tests and track contamination levels. Community-based monitoring helps identify contamination hotspots and supports targeted remediation efforts.

2. Advocating for Action

Communities play a vital role in advocating for PFAS management and remediation. Local advocacy groups can raise awareness, organise campaigns, and push for policy changes at the local and national levels. Effective advocacy ensures that PFAS issues are addressed promptly and comprehensively.

B. Supporting Health and Safety Measures

1. Providing Health Resources

Communities should support initiatives that offer health resources and screenings for individuals affected by PFAS contamination. Access to medical consultations, health assessments, and support services helps address the health impacts of PFAS exposure and provides necessary assistance to affected residents.

2. Implementing Safety Measures

Communities can implement safety measures to reduce PFAS exposure. This includes providing information on using water filters, avoiding contaminated food sources, and following guidelines for safe handling of potentially contaminated materials.

C. Promoting Sustainable Practices

1. Encouraging Sustainable Choices

Promoting sustainable practices within communities helps reduce the reliance on products that may contain PFAS. Community organisations can educate residents about sustainable alternatives and encourage practices that minimise environmental impact.

2. Participating in Recycling Programs

Communities should participate in and support recycling programs that prevent PFAS-containing products from entering landfills. Proper recycling and disposal help reduce the spread of PFAS and support overall environmental protection efforts.

4. Conclusion

Effective prevention and mitigation of PFAS contamination require a collaborative approach involving industries, policymakers, and communities. By adopting best practices, implementing robust strategies, and supporting proactive measures, stakeholders can address the risks associated with PFAS and protect public health and the environment. The recommendations outlined in this chapter offer a roadmap for developing and implementing effective prevention and mitigation strategies, contributing to a safer and healthier future.

Recommendations for Implementing Effective Prevention and Mitigation Measures

To effectively address PFAS contamination, it is crucial to adopt a structured approach that involves a combination of preventive measures, effective management strategies, and ongoing adaptation to emerging challenges. This part presents recommendations for implementing effective prevention and mitigation measures, focusing on the development of a national strategy, enhancements to regulatory frameworks, and the promotion of industry and community collaboration.

1. Developing a National PFAS Prevention Strategy

A. Formulating a Comprehensive Strategic Plan

**1. Establishing Clear Objectives

A national PFAS prevention strategy should outline specific, measurable objectives aimed at reducing PFAS emissions and contamination. These objectives could include reducing the use of PFAS in consumer products by a certain percentage, achieving significant reductions in PFAS levels in water sources, and eliminating PFAS discharges from key industries.

**2. Creating an Action Plan

The strategic plan should detail actionable steps to achieve the set objectives. This includes setting deadlines for implementation, identifying responsible parties, and allocating necessary resources. The action plan should address prevention, monitoring, and remediation efforts, ensuring a comprehensive approach to PFAS management.

B. Launching a National PFAS Task Force

**1. Composition and Responsibilities

A dedicated PFAS Task Force should be established to oversee the implementation of the national strategy. This task force should include representatives from government agencies, industry, academia, and community groups. Its responsibilities would include coordinating efforts, monitoring progress, and providing recommendations for policy adjustments.

**2. Regular Reporting and Accountability

The Task Force should be tasked with regular reporting on the progress of the PFAS prevention strategy. Transparent reporting ensures accountability and allows stakeholders to assess the effectiveness of measures. Regular updates also keep the public informed and engaged in PFAS management efforts.

2. Enhancing Regulatory Frameworks

A. Updating and Strengthening Regulations

**1. Implementing Stringent Limits and Standards

Regulations should be updated to include stringent limits for PFAS concentrations in air, water, and soil. Setting enforceable standards helps reduce the risk of widespread contamination and ensures that industries and other sources comply with safety limits.

**2. Establishing Comprehensive Guidelines

Comprehensive guidelines should be established for the use, disposal, and treatment of PFAS. These guidelines should cover various sectors, including industrial processes, firefighting operations, and consumer products. Clear guidelines help prevent PFAS releases and ensure consistent practices across different industries.

B. Enhancing Enforcement Mechanisms

**1. Strengthening Compliance Monitoring

Enforcement mechanisms should be enhanced to ensure compliance with PFAS regulations. This includes increasing the frequency of inspections, improving monitoring technologies, and implementing robust tracking systems for PFAS emissions and discharges.

**2. Imposing Penalties for Non-Compliance

Effective enforcement requires a system of penalties for non-compliance. Penalties should be significant enough to deter violations and incentivise adherence to regulations. This may include fines, operational restrictions, and legal actions against repeat offenders.

C. Supporting Research and Innovation

**1. Funding Research Initiatives

Policymakers should increase funding for research on PFAS detection, treatment, and alternatives. Research initiatives can lead to the development of innovative technologies and approaches for managing PFAS contamination.

**2. Promoting Collaboration and Knowledge Sharing

Encouraging collaboration between research institutions, industry, and government agencies can facilitate the exchange of knowledge and resources. Collaborative efforts can accelerate the development of effective solutions and enhance the overall understanding of PFAS issues.

3. Fostering Industry and Community Collaboration

A. Encouraging Industry Collaboration

**1. Promoting Industry Partnerships

Industries should be encouraged to collaborate on PFAS management initiatives. Partnerships between companies can lead to the sharing of best practices, development of joint solutions, and collective efforts to reduce PFAS use and emissions.

2. Implementing Industry-Wide Standards

Developing industry-wide standards for PFAS management can help ensure consistency and effectiveness. Industry associations and organisations can play a role in establishing these standards and promoting their adoption among members.

B. Supporting Community-Based Initiatives

1. Providing Resources for Local Efforts

Communities should be provided with resources to support local PFAS management efforts. This includes funding for environmental monitoring, educational programmes, and health screenings. Empowering communities with resources helps them effectively address contamination issues and advocate for necessary changes.

2. Facilitating Public Engagement

Public engagement is crucial for successful PFAS management. Communities should be involved in decision-making processes and given opportunities to provide input on local PFAS issues. Facilitating public engagement ensures that community concerns are addressed and that solutions are tailored to local needs.

4. Conclusion

Implementing effective prevention and mitigation measures for PFAS contamination requires a comprehensive and coordinated approach. By developing a national PFAS prevention strategy, enhancing regulatory frameworks, and fostering industry and community collaboration, stakeholders can address the risks associated with PFAS and work towards a safer and healthier environment. The recommendations outlined in this chapter provide a roadmap for achieving these goals and ensuring long-term success in managing PFAS contamination.

Recommendations for Implementing Effective Prevention and Mitigation Measures

Effective strategies for preventing and mitigating PFAS contamination require a multi-faceted approach, engaging various stakeholders and leveraging the latest technologies and methodologies. This final part provides a comprehensive set of recommendations for implementing these measures, focusing on practical steps that can be taken at different levels.

1. Strategic Recommendations for Industry

A. Implementing Best Practices

****1. Developing Internal PFAS Management Plans**

Industries should develop and implement internal PFAS management plans that outline strategies for minimising PFAS use, handling waste, and mitigating emissions. These plans should include:

- **Assessment of Current Use:** Identifying all processes and products that involve PFAS.
- **Substitution Strategies:** Finding and implementing safer alternatives.
- **Employee Training:** Ensuring staff are trained on PFAS handling and safety procedures.

****2. Investing in Research and Innovation**

****1. Supporting Innovation in Alternative Chemicals**

Industries should invest in research to discover and develop non-toxic alternatives to PFAS. Collaborating with research institutions can expedite this process and ensure that alternatives meet safety and performance standards.

****2. Implementing Pilot Projects**

Before wide-scale implementation, industries should conduct pilot projects to test the effectiveness of new PFAS-free technologies and practices. These trials help identify potential issues and refine solutions before broader application.

B. Enhancing Transparency and Communication
**1. Reporting and Disclosure

Industries should adhere to transparent reporting practices regarding PFAS use and emissions. Regular disclosure of PFAS data helps build trust with the public and regulators, and supports informed decision-making.

**2. Engaging with Stakeholders

Engaging with local communities, environmental groups, and regulators can facilitate collaboration and address concerns about PFAS pollution. Industry stakeholders should hold regular meetings to discuss issues, progress, and strategies for improvement.

2. Recommendations for Policymakers
A. Strengthening Legislation
**1. Updating Regulatory Standards

Policymakers should periodically review and update regulations to reflect the latest scientific knowledge and technological advancements. This includes setting lower allowable limits for PFAS in environmental media and consumer products.

**2. Enacting Proactive Legislation

Implementing proactive legislation that encourages the development of safer alternatives and supports research into PFAS remediation technologies can drive industry changes and improve public health outcomes.

B. Enhancing Enforcement and Compliance
**1. Increasing Inspection and Monitoring

Policymakers should allocate resources for more frequent inspections and monitoring of facilities that handle PFAS. Enhanced monitoring helps detect non-compliance early and ensures that mitigation measures are effectively implemented.

**2. Establishing Penalties and Incentives

Establishing a clear framework of penalties for non-compliance and incentives for exceeding regulatory requirements can motivate industries to adhere to best practices and adopt innovative solutions.

C. Supporting Public and Environmental Health

**1. Funding Health Impact Studies

Investing in studies that examine the health impacts of PFAS exposure is crucial for understanding the full extent of the problem. These studies can guide public health interventions and inform future regulations.

**2. Promoting Public Awareness Campaigns

Policymakers should support public awareness campaigns to educate communities about PFAS risks and prevention strategies. Effective communication helps individuals take appropriate actions to reduce exposure and advocate for policy changes.

3. Recommendations for Communities

A. Engaging in Local Monitoring and Advocacy

**1. Forming Community Watch Groups

Communities should establish local watch groups to monitor PFAS contamination and advocate for environmental protection. These groups can collaborate with local authorities and researchers to address contamination issues and ensure accountability.

**2. Organising Public Meetings and Forums

Public meetings and forums provide a platform for discussing PFAS-related issues, sharing information, and mobilising community action. These gatherings can also serve as a venue for expressing concerns to policymakers and industry representatives.

B. Implementing Personal and Household Measures

**1. Using Water Filtration Systems

Households in affected areas should consider installing water filtration systems designed to remove PFAS from drinking water. High-quality filters can significantly reduce PFAS levels and protect health.

****2. Avoiding Contaminated Products**

Individuals should be aware of potential PFAS-containing products and avoid them when possible. This includes products with stain-resistant coatings, non-stick cookware, and certain types of packaging.

C. Promoting Environmental Stewardship

****1. Participating in Cleanup Efforts**

Communities can participate in or organise cleanup efforts to address PFAS contamination in local environments. These activities may include river cleanups, soil remediation projects, and waste disposal initiatives.

****2. Supporting Sustainable Practices**

Promoting and adopting sustainable practices within the community can help reduce reliance on PFAS-containing products. This includes supporting businesses that use environmentally friendly materials and advocating for green practices.

4. Conclusion

Implementing effective prevention and mitigation measures for PFAS contamination requires a concerted effort from industries, policymakers, and communities. By following the recommendations outlined in this chapter, stakeholders can take meaningful actions to address PFAS issues, protect public health, and safeguard the environment. A comprehensive and coordinated approach will help ensure long-term success in managing PFAS contamination and achieving a healthier, safer future.

A Call to Action – Moving Forward

Summary of Key Findings and Recommendations

As we reach the concluding chapter of this comprehensive examination of PFAS contamination, it is imperative to synthesise the critical findings and actionable recommendations from the previous chapters. This summary will outline the key insights gained through the exploration of PFAS pollution, the current state of UK waters, health and ecological impacts, regulatory frameworks, industry roles, and more. The goal is to provide a clear and concise overview of the issues at hand and present a unified call to action for all stakeholders involved.

1. Understanding PFAS and Their Impacts

A. Nature and Persistence of PFAS

Per- and polyfluoroalkyl substances (PFAS) are a group of synthetic chemicals known for their resistance to heat, water, and oil. Their chemical stability, coupled with their widespread use in various products—from firefighting foams to non-stick cookware—has resulted in significant environmental and health challenges. PFAS are often referred to as "forever chemicals" due to their persistence in the environment and human body.

B. Health Risks Associated with PFAS

Research has established a strong link between PFAS exposure and a range of adverse health outcomes, including cancer, liver damage, and immune system dysfunction. Epidemiological studies have consistently shown elevated risks of these health issues among populations with higher levels of PFAS exposure, reinforcing the urgency of addressing this contamination.

C. Ecological Consequences

The ecological impact of PFAS contamination is profound, affecting aquatic life and ecosystems. PFAS bioaccumulate in fish and other wildlife, leading to disruptions in biodiversity and ecosystem function. Long-term consequences may include altered food chains, reduced species diversity, and compromised ecosystem health.

2. Current Contamination Levels and Regulatory Efforts

A. State of UK Waters

Recent data indicates that PFAS contamination is a significant issue in UK waters, including rivers, lakes, and groundwater. Monitoring efforts have revealed concerning levels of PFAS in these sources, underscoring the need for more robust testing and management practices.

B. Regulatory Frameworks

The UK's regulatory approach to PFAS has evolved, but challenges remain. Existing regulations and guidelines are in place to manage PFAS contamination, but there is room for improvement in enforcement and adaptation to new scientific insights. Comparing the UK's policies with those of other countries reveals both strengths and areas needing enhancement.

3. Industry Roles and Technological Advances

A. Industry Responsibility

Industries play a crucial role in PFAS contamination, both as major users of these chemicals and as contributors to environmental discharges. While some companies have taken steps to address PFAS issues, there remains a need for broader industry-wide commitments to reduce PFAS use and improve waste management practices.

B. Technological Innovations

Advancements in PFAS detection and remediation technologies offer hope for better management of contamination. Innovations in filtration, destruction, and containment techniques are critical for addressing existing pollution and preventing future contamination.

4. Public Awareness and Advocacy

A. Raising Public Awareness

Public awareness of PFAS contamination and its health impacts is vital for driving action and supporting regulatory changes. Environmental NGOs, media, and community groups have played essential roles in raising awareness and advocating for stricter regulations and better management practices.

B. Case Studies and Advocacy Successes

Case studies of successful advocacy efforts illustrate the power of public and community action in influencing policy changes and improving environmental protections. These examples serve as models for future advocacy and highlight the importance of ongoing engagement and vigilance.

5. Recommendations for Moving Forward

A. Developing and Implementing Comprehensive Strategies

1. **National PFAS Prevention Strategy:** Establish a cohesive national strategy that includes clear objectives, action plans, and a dedicated task force to coordinate efforts and monitor progress.

2. **Enhanced Regulations:** Update and strengthen regulatory frameworks to include stricter limits on PFAS concentrations, comprehensive guidelines, and robust enforcement mechanisms.

3. **Industry and Community Collaboration:** Foster partnerships between industries, communities, and policymakers to develop and implement best practices, share knowledge, and support local initiatives.

B. Investing in Research and Innovation

1. **Supporting Research:** Increase funding for research on PFAS detection, alternatives, and remediation technologies to drive innovation and improve management practices.

2. **Promoting Collaboration:** Encourage collaboration between research institutions, industry, and government agencies to accelerate the development and adoption of effective solutions.

C. Engaging Stakeholders

1. **Public Engagement:** Enhance public awareness through education and advocacy efforts to empower individuals and communities to take action and support PFAS management initiatives.
2. **Policy Advocacy:** Advocate for policy changes at the local, national, and international levels to address PFAS contamination comprehensively and effectively.

6. Vision for the Future

The vision for a future with improved PFAS management and reduced contamination in UK waters involves a concerted effort from all stakeholders. By implementing the recommendations outlined in this chapter, we can achieve the following goals:

A. Reduced PFAS Levels

Through effective prevention, stringent regulations, and technological advancements, PFAS levels in water sources can be significantly reduced, leading to improved environmental and public health outcomes.

B. Enhanced Public Health

By addressing PFAS contamination, we can mitigate the health risks associated with these chemicals, ensuring a healthier population and reducing the burden of PFAS-related diseases.

C. Sustainable Practices

Promoting sustainable practices and alternatives to PFAS will help prevent future contamination and support a cleaner, safer environment for current and future generations.

D. Informed and Engaged Communities

Empowered communities, equipped with knowledge and resources, will play a vital role in driving change and ensuring that PFAS issues are addressed effectively.

7. Conclusion

In conclusion, the call to action for PFAS management is clear. It requires a unified effort from government agencies, industries, and the public to address the challenges posed by PFAS contamination. By taking the steps outlined in this chapter, stakeholders can contribute to a safer, healthier future, free from the pervasive impact of "forever chemicals." The journey towards improved PFAS management begins with collective action and a commitment to making meaningful changes for the benefit of all.

A Call to Action for Stakeholders

As we move forward, it is imperative that all stakeholders—government agencies, industries, and the public—take proactive steps to address PFAS contamination. Each group has a unique role and responsibility in tackling this complex issue. This section outlines a call to action tailored for each stakeholder, aiming to foster collaboration and drive effective solutions.

1. Call to Action for Government Agencies

A. Strengthening Policy and Legislation

**1. Enhance Regulatory Frameworks

Government agencies must prioritise updating and strengthening regulatory frameworks to address PFAS contamination comprehensively. This includes:

- **Implementing Stricter Standards:** Develop and enforce stricter standards for PFAS levels in water, soil, and air. Set lower permissible limits based on the latest scientific research to protect public health and the environment.

- **Expanding Monitoring Requirements:** Increase the frequency and scope of monitoring for PFAS in water sources, soils, and consumer products. Ensure that data is transparently reported and accessible to the public.

****2. Promote Research and Innovation**

- **Invest in Research:** Allocate funding to research institutions focused on PFAS detection, alternatives, and remediation technologies. Encourage innovation by supporting grants and collaborative projects.
- **Support Technological Development:** Facilitate the development and implementation of advanced technologies for PFAS removal and containment. Foster partnerships between research institutions, technology developers, and regulatory bodies.

B. Enhancing Enforcement and Compliance
****1. Increase Enforcement Activities**

- **Conduct Regular Inspections:** Implement rigorous inspection regimes for industries handling PFAS. Ensure compliance with regulations through frequent checks and prompt corrective actions for non-compliance.
- **Impose Penalties:** Develop a robust system of penalties for violations of PFAS regulations. Ensure that fines and sanctions are significant enough to deter non-compliance and encourage adherence to best practices.

****2. Strengthen Coordination and Collaboration**

- **Facilitate Interagency Cooperation:** Enhance

coordination between various government departments and agencies involved in environmental protection, public health, and industrial regulation. A unified approach ensures comprehensive management of PFAS contamination.

- **Engage with International Bodies:** Collaborate with international organisations and other countries to share knowledge, best practices, and strategies for PFAS management. Adopt and adapt successful approaches from other jurisdictions.

2. Call to Action for Industries
A. Implementing Best Practices
**1. Adopt Safer Alternatives

- **Identify Alternatives:** Evaluate and adopt safer alternatives to PFAS in products and processes. Invest in research to find and implement substitutes that do not pose similar environmental and health risks.
- **Enhance Product Safety:** Ensure that products marketed as PFAS-free are tested and verified through independent assessments. Provide clear information to consumers about the absence of harmful chemicals.

**2. Improve Waste Management

- **Implement Waste Reduction Strategies:** Develop and implement strategies to minimise PFAS waste at the source. Improve waste handling and disposal practices to prevent environmental contamination.
- **Invest in Treatment Technologies:** Invest in advanced treatment technologies for PFAS-containing waste. Ensure

that waste treatment facilities are equipped to handle and neutralise PFAS effectively.

B. Promoting Transparency and Accountability
**1. Increase Reporting and Disclosure

- **Disclose PFAS Use:** Regularly report PFAS usage, emissions, and management practices. Provide detailed information on PFAS concentrations in products and processes to regulatory bodies and the public.
- **Engage in Public Reporting:** Publish environmental impact reports and updates on PFAS management efforts. Transparency builds trust and demonstrates commitment to addressing contamination.

**2. Engage with Stakeholders

- **Collaborate with Communities:** Engage with local communities affected by PFAS contamination. Address concerns, provide information, and work together to develop and implement solutions.
- **Participate in Industry Forums:** Join industry forums and working groups focused on PFAS management. Share knowledge and collaborate on initiatives to advance best practices and solutions.

3. Call to Action for the Public
A. Increasing Awareness and Advocacy
**1. Educate and Inform

- **Raise Awareness:** Educate yourself and your community about the risks of PFAS contamination and how to minimise exposure. Participate in public awareness campaigns and informational sessions.

- **Advocate for Change:** Support and advocate for policies and regulations aimed at reducing PFAS contamination. Engage with local and national campaigns to push for stronger environmental protections.

**2. Engage in Community Action

- **Participate in Local Initiatives:** Get involved in local initiatives and organisations working on PFAS issues. Volunteer for cleanup efforts, participate in monitoring programmes, and support local advocacy groups.
- **Organise Community Events:** Host or attend community events focused on PFAS education and action. Use these platforms to discuss concerns, share information, and mobilise collective action.

B. Implementing Personal Precautions
**1. Reduce Personal Exposure

- **Use Filtration Systems:** Consider installing water filtration systems that are effective at removing PFAS from drinking water. Ensure that filters are regularly maintained and replaced as needed.
- **Avoid PFAS-Containing Products:** Be cautious when purchasing products with stain-resistant or water-repellent coatings. Opt for PFAS-free alternatives whenever possible.

**2. Support Sustainable Practices

- **Adopt Eco-Friendly Choices:** Support businesses and products that prioritise sustainability and avoid PFAS. Choose environmentally friendly options in your daily life and encourage others to do the same.
- **Promote Environmental Stewardship:** Advocate for and practice responsible environmental stewardship. Participate in or support initiatives aimed at reducing pollution and protecting natural resources.

4. Conclusion

The call to action for PFAS management is a collective responsibility that requires commitment and collaboration from all stakeholders. Government agencies, industries, and the public must work together to address the challenges posed by PFAS contamination effectively. By implementing the recommendations outlined in this chapter, we can achieve meaningful progress in reducing PFAS pollution, protecting public health, and safeguarding the environment. The journey towards a future free from the pervasive impact of PFAS begins with concerted action and a shared dedication to creating a cleaner, safer world for all.

Vision for the Future of PFAS Management

As we confront the ongoing challenge of PFAS contamination, envisioning a future where effective management and mitigation are the norms rather than the exceptions is crucial. This vision involves a multi-faceted approach integrating advanced technologies, robust policies, proactive public engagement, and collaborative efforts across all sectors. This section will outline the aspirations and strategies for a future with improved PFAS management, focusing on key areas of progress and innovation.

1. Advancing Technological Solutions

A. Innovations in Detection and Monitoring

**1. Enhanced Detection Methods

The future of PFAS management hinges on the development and deployment of advanced detection technologies. Current methods, while effective, must evolve to provide more sensitive, accurate, and cost-effective means of identifying PFAS contaminants in various media.

- **High-Resolution Mass Spectrometry:** Advances in mass spectrometry can improve the detection of PFAS at lower concentrations, allowing for earlier identification and more comprehensive monitoring.
- **Remote Sensing Technologies:** Implementing remote sensing technologies, such as satellite imagery and aerial surveys, could provide broader and more detailed data on PFAS contamination across large areas.

**2. Real-Time Monitoring Systems

- **Continuous Monitoring:** The integration of real-time monitoring systems in water treatment facilities and environmental sensors can offer immediate feedback on PFAS levels. This capability would enable prompt responses to contamination events and better management of water quality.
- **Data Analytics:** Leveraging big data analytics and artificial intelligence (AI) can enhance the interpretation of monitoring data, identifying trends and predicting potential contamination hotspots with greater precision.

B. Remediation Technologies
1. Innovative Remediation Methods
Effective PFAS remediation requires ongoing innovation

to develop methods that are both efficient and cost-effective. The future will likely see a shift towards more advanced technologies capable of addressing PFAS contamination at various scales.

- **Advanced Oxidation Processes (AOPs):** Techniques such as photocatalysis and electrochemical oxidation show promise in breaking down PFAS compounds into less harmful substances. These methods can potentially be integrated into existing water treatment systems.
- **Bioremediation:** Research into microbial and enzymatic solutions for PFAS degradation is advancing. Specific bacteria and enzymes that can metabolise PFAS are being identified, offering a natural and potentially scalable method for cleanup.

****2. Sustainable Solutions**

- **Green Chemistry:** The development of green chemistry principles can lead to the creation of safer chemical alternatives that do not pose similar environmental and health risks as PFAS. These alternatives can be designed to avoid the persistence and toxicity associated with traditional PFAS.
- **Circular Economy Practices:** Embracing circular economy principles—such as recycling and reusing materials—can help reduce the production and disposal of PFAS-containing products, thereby minimising the overall environmental footprint.

2. Strengthening Regulatory Frameworks
A. Implementing Robust Regulations
****1. Stricter Standards and Limits**

The establishment of stringent regulations is essential for effective PFAS management. Regulatory bodies must:

- **Set Lower Thresholds:** Adopt more stringent limits for PFAS concentrations in drinking water, soil, and air based on the latest scientific research and risk assessments.
- **Comprehensive Regulations:** Develop regulations that cover all sources of PFAS contamination, including industrial processes, consumer products, and waste disposal.

2. Enhanced Enforcement and Compliance

- **Regulatory Enforcement:** Strengthen enforcement mechanisms to ensure compliance with PFAS regulations. This includes conducting regular inspections, increasing penalties for violations, and promoting transparency in reporting.
- **Public Participation:** Encourage public involvement in regulatory processes by providing opportunities for community input and feedback on PFAS management policies.

B. International Collaboration
1. Global Standards and Agreements

PFAS contamination is a global issue requiring international cooperation. Future efforts should include:

- **International Agreements:** Support and participate in international agreements and treaties aimed at reducing PFAS emissions and managing contamination on a global scale.

- **Knowledge Sharing:** Collaborate with international organisations to share knowledge, best practices, and technological advancements in PFAS management.

2. Cross-Border Initiatives

- **Regional Cooperation:** Engage in regional cooperation efforts to address PFAS contamination in shared water bodies and ecosystems. Joint initiatives can enhance the effectiveness of management strategies and improve resource allocation.

3. Empowering Public and Community Engagement
A. Raising Awareness and Education
1. Educational Campaigns

Public education and awareness are critical to driving change and fostering a culture of environmental stewardship.

- **Community Workshops:** Organise workshops and informational sessions to educate communities about PFAS risks, prevention measures, and available resources.
- **Public Outreach:** Use media, social platforms, and community events to raise awareness about PFAS contamination and encourage public involvement in advocacy and mitigation efforts.

2. Support for Affected Communities

- **Resources and Assistance:** Provide resources and support to communities affected by PFAS contamination, including health screenings, legal assistance, and financial aid for remediation efforts.
- **Community Action Plans:** Facilitate the development of community action plans that outline local strategies for

addressing PFAS contamination and engaging residents in the process.

B. Fostering Collaboration and Advocacy
**1. Building Partnerships

Collaboration among stakeholders—government agencies, industries, non-governmental organisations, and the public—is essential for effective PFAS management.

- **Multi-Stakeholder Initiatives:** Create platforms for dialogue and collaboration among diverse stakeholders to address PFAS issues collectively and develop comprehensive solutions.
- **Support for Advocacy Groups:** Support and partner with advocacy groups working to raise awareness, drive policy changes, and promote best practices in PFAS management.

**2. Encouraging Civic Engagement

- **Citizen Science:** Encourage citizen science projects that involve community members in monitoring and reporting PFAS contamination. This can enhance data collection efforts and empower individuals to take an active role in environmental protection.
- **Policy Advocacy:** Mobilise communities to advocate for stronger PFAS regulations and support legislative initiatives aimed at improving environmental and public health protections.

4. Preparing for the Future
A. Anticipating Emerging Challenges

As we move forward, it is crucial to anticipate and prepare for emerging challenges related to PFAS contamination.

- **New PFAS Compounds:** Monitor and assess the potential risks posed by emerging PFAS compounds and their degradation products. Ensure that regulatory frameworks and remediation technologies are adaptable to address new challenges.
- **Long-Term Monitoring:** Establish long-term monitoring programmes to track the effectiveness of PFAS management strategies and identify areas for improvement.

B. Vision for a Cleaner Future

The vision for a future free from PFAS contamination involves a collaborative effort to reduce emissions, implement effective management practices, and safeguard public health and the environment. By embracing innovation, strengthening regulations, and fostering community engagement, we can achieve a cleaner, safer world.

Conclusion

The future of PFAS management is within our reach if we commit to taking decisive and collaborative action. Through advancements in technology, robust regulatory frameworks, public engagement, and international cooperation, we can create a future where PFAS contamination is effectively managed and its impact on health and the environment is significantly reduced. The journey towards this future begins with each stakeholder playing their part in the collective effort to protect our waters, our health, and our planet.

Vision for a Future with Improved PFAS Management

To envision a future with improved PFAS management, we must integrate forward-thinking strategies across various dimensions of policy, technology, community involvement, and international cooperation. The goal is not merely to mitigate existing contamination but to prevent future occurrences and foster a safer, healthier environment. This section explores the comprehensive strategies needed to achieve these objectives, laying out a roadmap for effective PFAS management in the years to come.

1. Establishing a Comprehensive Management Framework

A. Developing Integrated Policy Approaches

****1. Holistic Policy Framework**

A successful future approach to PFAS management involves developing an integrated policy framework that addresses all aspects of PFAS contamination—from prevention and detection to remediation and recovery.

- **National Strategy:** Formulate a national PFAS strategy that aligns with public health goals, environmental protection standards, and economic considerations. This strategy should be flexible, allowing for updates as new information and technologies become available.

- **Localised Plans:** Develop regional and local action plans tailored to specific contamination issues and community needs. These plans should be coordinated with national policies to ensure a cohesive approach.

****2. Incorporating Public Health Perspectives**

- **Health Impact Assessments:** Integrate health impact assessments into policy development to understand and mitigate the potential health effects of PFAS exposure. Use these assessments to inform regulatory standards and public health interventions.

- **Community Health Initiatives:** Support public health initiatives aimed at reducing exposure to PFAS and addressing health issues related to contamination. This includes funding for medical research, health screenings, and community health programmes.

B. Advancing Technological Innovations
**1. Accelerating Research and Development
**1. Funding and Support

To drive innovation in PFAS management, it is essential to invest in research and development efforts.

- **Government Grants:** Increase funding for research initiatives focused on PFAS detection, remediation, and alternatives. Support collaborations between academia, industry, and government agencies.
- **Public-Private Partnerships:** Foster partnerships between public entities and private companies to accelerate the development and deployment of new technologies. These partnerships can leverage resources and expertise to advance solutions more effectively.

**2. Promoting Technological Adoption

- **Technology Deployment:** Ensure the widespread adoption of advanced technologies for PFAS detection and treatment. Provide incentives for industries and municipalities to implement these technologies in their operations and facilities.
- **Training and Capacity Building:** Offer training programmes for professionals involved in PFAS management to ensure they are equipped with the knowledge and skills to utilise new technologies effectively.

2. Strengthening Global Collaboration
A. Enhancing International Cooperation
****1. Global Agreements and Initiatives**

To tackle the global nature of PFAS contamination, robust international cooperation is necessary.

- **International Treaties:** Support and engage in international treaties and agreements aimed at reducing PFAS emissions and managing contamination. Work towards global commitments to phase out the use of hazardous PFAS compounds.
- **Global Research Networks:** Participate in global research networks that focus on PFAS issues. Share data, research findings, and best practices to enhance collective understanding and response strategies.

****2. Cross-Border Projects**

- **Transboundary Initiatives:** Collaborate on cross-border projects to address PFAS contamination in shared water bodies and ecosystems. Joint efforts can improve the effectiveness of remediation and prevention measures.
- **Regional Cooperation:** Engage in regional forums and collaborations to address PFAS issues specific to geographical areas. Develop coordinated action plans and resource-sharing agreements.

B. Fostering Local and Community Engagement
****1. Empowering Local Action**

Local communities play a crucial role in PFAS management, and empowering them can lead to effective and sustainable solutions.

- **Community-Based Projects:** Support community-led projects focused on PFAS monitoring, education, and remediation. Provide resources and guidance to help communities implement local initiatives.
- **Local Advocacy:** Encourage local advocacy efforts to raise awareness and push for stronger regulations and policies. Empower residents to be actively involved in decision-making processes related to PFAS management.

2. Building Resilient Communities

- **Community Resilience Planning:** Develop resilience plans that address the social, economic, and environmental impacts of PFAS contamination. Ensure that communities are prepared to respond to and recover from contamination events.
- **Support Networks:** Create support networks for affected communities, including access to health services, legal assistance, and financial aid for remediation efforts.

3. Promoting Sustainable Practices
A. Encouraging Sustainable Industrial Practices
1. Adopting Green Technologies
Industries have a critical role in reducing PFAS contamination through the adoption of sustainable practices.

- **Green Chemistry:** Promote the use of green chemistry principles to develop safer alternatives to PFAS. Encourage industries to shift towards non-toxic materials and processes.
- **Sustainable Manufacturing:** Support manufacturing practices that minimise the generation of PFAS waste and

ensure proper disposal and treatment of PFAS-containing products.

2. Corporate Social Responsibility

- **Transparency and Reporting:** Encourage companies to be transparent about their PFAS use and emissions. Implement robust reporting mechanisms and public disclosures to hold industries accountable.
- **Community Engagement:** Foster corporate social responsibility by engaging with communities affected by PFAS contamination. Contribute to remediation efforts and support local initiatives aimed at reducing exposure.

B. Promoting Eco-Friendly Consumer Choices
1. Consumer Education

Educating consumers about PFAS risks and safer alternatives is vital for reducing personal exposure and driving market demand for non-toxic products.

- **Product Labelling:** Advocate for clear labelling of products to indicate the absence of PFAS and other harmful chemicals. Provide consumers with information to make informed choices.
- **Public Awareness Campaigns:** Launch campaigns to raise awareness about PFAS and promote eco-friendly products. Encourage consumers to support businesses that prioritise sustainability and safety.

2. Supporting Environmental Stewardship

- **Environmental Advocacy:** Engage in environmental

stewardship efforts to protect water sources and ecosystems from PFAS contamination. Support initiatives aimed at preserving natural resources and promoting sustainable practices.

- **Volunteer and Community Activities:** Participate in or organise volunteer activities focused on environmental protection and PFAS mitigation. Collaborate with local organisations to support community-based environmental efforts.

4. Conclusion

The vision for a future with improved PFAS management involves a comprehensive approach that integrates advanced technologies, robust policies, global collaboration, local engagement, and sustainable practices. By embracing these strategies, we can work towards a future where PFAS contamination is effectively managed, public health is safeguarded, and environmental quality is preserved. The collective efforts of government agencies, industries, communities, and individuals will be crucial in achieving this vision and ensuring a safer, cleaner world for future generations. The path forward is challenging, but with determination and cooperation, we can make significant strides towards addressing the PFAS crisis and creating a sustainable future.

Don't miss out!

Visit the website below and you can sign up to receive emails whenever Harper Nomad publishes a new book. There's no charge and no obligation.

https://books2read.com/r/B-A-EDIOB-BQDWD

BOOKS 2 READ

Connecting independent readers to independent writers.

Also by Harper Nomad

Bible Deep Dive
Journey of Faith: Lessons From the Book of Genesis
Journey of Faith: Lessons from the Book of Leviticus
Journey of Faith: Lessons from the Book of Numbers
Journey of Faith: Lessons from the Book of Deuteronomy
Journey of Faith: Lessons from the Book of Joshua

Standalone
Mastering Social Media for Authors
Journey of Faith: Lessons from the Book of Exodus
Journey Through Ancient Rome: An Immersive Travel Guide
Unlock Financial Freedom
Discovering Torrequebrada: A Pocket Tourist Guide
Sour Water: Should We Be Worried?